Methodology of Educational Measurement and Assessment

Series editors

[1]This work was conducted while M. von Davier was employed with Educational Testing Service.

This new book series collates key contributions to a fast-developing field of education research. It is an international forum for theoretical and empirical studies exploring new and existing methods of collecting, analyzing, and reporting data from educational measurements and assessments. Covering a high-profile topic from multiple viewpoints, it aims to foster a broader understanding of fresh developments as innovative software tools and new concepts such as competency models and skills diagnosis continue to gain traction in educational institutions around the world. Methodology of Educational Measurement and Assessment offers readers reliable critical evaluations, reviews and comparisons of existing methodologies alongside authoritative analysis and commentary on new and emerging approaches. It will showcase empirical research on applications, examine issues such as reliability, validity, and comparability, and help keep readers up to speed on developments in statistical modeling approaches. The fully peer-reviewed publications in the series cover measurement and assessment at all levels of education and feature work by academics and education professionals from around the world. Providing an authoritative central clearing-house for research in a core sector in education, the series forms a major contribution to the international literature.

More information about this series at http://www.springer.com/series/13206

Jorge González • Marie Wiberg

Applying Test Equating Methods

Using R

Springer

Jorge González
Faculty of Mathematics
Pontificia Universidad Católica de Chile
Santiago, Chile

Marie Wiberg
Department of Statistics
Umeå School of Business and Economics
Umeå University
Umeå, Sweden

ISSN 2367-170X ISSN 2367-1718 (electronic)
Methodology of Educational Measurement and Assessment
ISBN 978-3-319-51822-0 ISBN 978-3-319-51824-4 (eBook)
DOI 10.1007/978-3-319-51824-4

Library of Congress Control Number: 2017932073

Printed on acid-free paper

This Springer imprint is published by Springer Nature
The registered company is Springer International Publishing AG
The registered company address is: Gewerbestrasse 11, 6330 Cham, Switzerland

To Javiera and Renato

– J.G.

To Emelie and Hanna

– M.W.

Foreword

This book signals the maturity of the methodologies and technology that surround test score equating and the acceptance of these methodologies and of the open-source software by the testing industry's experts and academia.

Test equating is a statistical procedure for adjusting the test form differences in standardized testing. Test equating methodologies originated in testing companies many decades ago. Due to the pragmatic operational perspective, many of these methodologies had lacked rigorous theoretical background and research for many years. Then in the 1980s, test score equating started to receive the appropriate attention from researchers and academia (Holland and Rubin 1982). Several notable journal articles and books were written at that time, and the field was brought forward significantly. The availability of personal and fast computers also facilitated the implementation of the established equating methods. For example, psychometricians and technology specialists at Educational Testing Service developed **GENASYS**, a very comprehensive, modular, and sophisticated software system to support the large-scale application of test score analysis and equating with all the available methodologies at that time. The first edition of the Kolen and Brennan (1995) book became a landmark in equating. A second wave of theoretical interest and the development of new methods took place around 2005 (Kolen and Brennan 2004; von Davier et al. 2004; Dorans et al. 2007).

Since 2005, there has been a noticeable shift in test equating. First, there has been an increasing interest in and use of test equating all over the world. Second, as it was illustrated in the edited volume of von Davier (2011a), there has been a renewed interest in searching for better methodologies that improve the score conversion in terms of bias (the observed-score equating framework, various kernel methods, local equating), in terms of error (kernel equating, equating with exponential families), in terms of hypothesis testing (Lagrange multiplier tests for item response theory (IRT) methods, Wald test for kernel methods), in terms of approximations for small sample equating, and in terms of monitoring the quality control of equating results (using time series approaches); third, there has been an increased interest in equating by university professors in statistics and educational measurement; and more importantly, there has been a significant software development using different software environments, from SAS to **R**. Moreover, the rapid changes in the technology made the upgrade of proprietary software challenging; the society became more appreciative of the open-source environment that facilitates transparency, research, and community work.

This volume is remarkable from several perspectives. First, it is a perfect mirror of the changes in test equating during recent years: it includes new equating methods alongside the older ones, and it provides code in **R** for users. The authors of this volume have both an academic background (one works for the Pontificia Universidad Católica de Chile, in Santiago de Chile, and one works for Umeå University in Sweden) and a practitioner background (one is a consultant for the operational tests from the measurement center MIDE UC and the other consults for the operational large-scale Swedish SAT). Both authors developed or codeveloped R packages for conducting various equating methods. Last but not least, one is from Chile and one is from Sweden. Again, the technology makes the world interconnected and supports collaborations like these!

The structure of this volume follows the process of operationalizing various test equating methods. The volume starts by locating equating in the realm of statistical models, which was also attempted by von Davier (2011b). Then the volume continues by briefly presenting the theory behind the methods (smoothing of discrete test score data, linear interpolation equating, several kernel methods, local and IRT equating) and by illustrating the application of these methods with sets of real data and with various **R** packages, including those developed by the authors themselves.

For those of us who work on educational assessments, there are still many open challenges. Measurement accuracy for all groups of test takers across the range of scores is hard to ensure in operational testing. With the increased shifts in demographics, we had to provide more sophisticated and complex, yet transparent, types of tests. Nevertheless, as we test more diverse groups of people, the challenge of providing accurate scores increases. For example, much of the research on alternative kernels, in particular on adaptive kernels, has been motivated by the need to analyze data from heterogeneous populations. Another way to look at fairness is through the lenses of transparency of the process: Will the transparency of the open-source software aid the fairness arguments?

Another area that remains of interest for future research is how to ensure that the computations with the open-source software environment are well tested and error-free for large-scale high-stakes tests administered over time. In the future, we will also need to provide open-source code for methods for quality control.

As many of us have started to work on simulation- and game-based assessments, new challenges occurred. To meet these challenges, we need to redefine the concept of test equating and the notion of adjusting for difficulty across test forms. While we do not have a solution yet, it is quite probable that the solution will use some of these newer approaches to test equating and open-source software.

A final challenge relates to improved communication about the practical consequences of addressing equating issues as an ongoing activity in order to ensure meaningful scores. We need to explain the limitations of the methods and the cost of being able to make truthful claims about score accuracy for all groups over time. The authors of this book examine theoretical and practical issues that cut across different types of methodological and operational situations and open the doors to hands-on teaching and learning and to more theoretical research on equating methods conducted by students and other researchers.

Princeton, USA
December 2016

Alina A. von Davier

References

Dorans, N. J., Pommerich, M., & Holland, P. W. (2007). *Linking and aligning scores and scales*. New York: Springer.

Holland, P., & Rubin, D. (1982). *Test equating*. New York: Academic Press.

Kolen, M., & Brennan, R. (1995). *Test equating: Methods and practices* (1st ed.). New York: Springer.

Kolen, M., & Brennan, R. (2004). *Test equating, scaling, and linking: Methods and practices* (2nd ed.). New York: Springer.

von Davier, A. (2011a). *Statistical models for test equating, scaling, and linking*. New York: Springer.

von Davier, A. (2011b). A statistical perspective on equating test scores. In A. von Davier (Ed.), *Statistical models for test equating, scaling, and linking* (Vol. 1, pp. 1–17). New York: Springer.

von Davier, A. A., Holland, P., & Thayer, D. (2004). *The kernel method of test equating*. New York: Springer.

Preface

This book provides an overview of test equating methods with a focus on how to implement them using **R** (R Core Team 2016).

The idea for writing this book emerged in April 2014 when we both attended the NCME conference in Philadelphia. There are many reasons to go to conferences including presenting your research, meeting other researchers, working on collaborative projects, and getting new ideas. For us, writing a book on equating with applications using **R** was a natural step as we had both been involved in the development of two different **R** packages, **kequate** (Andersson et al. 2013) and **SNSequate** (González 2014), and we have noticed that a number of software and **R** packages that can be used when equating test scores have emerged in recent years. Weeks (2010) (and references therein) offers a list that includes **EQUATE, ST, mcmequate, IpLink, POLYST, STUIRT, POLYEQUATE, IRTEQ, irtoys**, and **MischPsycho**. These software packages are mainly designed for conducting item response theory (IRT) linking rather than equating, with the exception of **POLYEQUATE** and **IRTEQ** that implement IRT equating. The list can be enlarged by including **CIPE, Equating Recipes, RAGE-RGEQUATE, Equating Error, PIE, and LEGS** (Kolen and Brennan 2014), the **KE** software (ETS 2011), the SAS/IML macro for log-linear smoothing (Moses and von Davier 2011), and the **R** packages **plink** (Weeks 2010), **lordif** (Choi et al. 2011), **equate** (Albano 2016), and **equateIRT** (Battauz 2015).

Two questions of interest to us were whether all of the available equating methods had been implemented in these packages and if we could replicate known results using all of these different packages. When we started writing this book, neither of us had used all of these packages, but we got to learn many things and got many ideas for new research along the way. By using an open-source software, a new generation can easily start their research where the previous generation left off. Thus, we expect many new research papers within the equating field in the next few years.

People in the test equating field might wonder what is new and unique in this equating book compared with others such as the works of von Davier et al. (2004) and Kolen and Brennan (2014). This is the first equating book that provides all of the necessary tools to perform equating regardless of which framework the equating belongs to. This means that no matter if it is a traditional equating method, an IRT equating method, a local equating method, or a kernel equating method, this book provides the essential tools both in terms of theory and computational implementation. The goal is to give the readers a pamphlet of as many different equating methods as possible and to provide them with hands-on examples of how to perform the analyses by themselves. A unique feature of this book is that it provides **R** codes to implement all of the described equating methods. **R** has become an essential tool for research in many fields including psychometrics, and our aim is to ensure that equating methods are not the exception of this tendency.

The intended audience for this book is very broad. It includes test constructors, researchers in equating, graduate students both in statistics and psychometrics, and all of those who have an interest in learning how to perform equating with different methods. One of the aims of the book is also to make it easier for beginners to get into the research field of equating.

Organization of this book. The first chapter gives a summary of our view of the equating transformation as a statistical estimator. It then moves on to the practical concerns in Chap. 2 in which the data used in the book are described. This chapter also explains how the data need to be prepared and structured in order to be utilized with the different **R** packages used in the book. Chapter 3 contains descriptions and illustrations of the traditional equating methods, while Chap. 4 goes step by step through the kernel equating framework. Chapter 5 describes and illustrates different IRT equating methods, and Chap. 6 contains material regarding some of the existing local equating methods. Finally, in Chap. 7, we illustrate how to replicate some recent developments in test equating research. In all chapters, examples using **R** code are provided. All used data sets and **R** scripts as well as additional examples not appearing in the book are available from the book's Web page at http://www.mat.uc.cl/~jorge.gonzalez/EquatingRbook.

Jorge's Acknowledgments. Writing this book has been possible thanks to many people who have been important in my career. Although my Ph.D. dissertation was not related to equating, I was lucky to be part of the psychometric research group at K.U. Leuven and to have Paul de Boeck and Francis Tuerlinckx as supervisors. Once I graduated, I was hired by the Measurement Center MIDE UC where I learned many things related to educational measurement, equating being among them. I thank Jorge Manzi, director of the Measurement Center MIDE UC, for trusting me and allowing me to be part of the center. I have also been very lucky to meet wonderful colleagues and friends from whom I have learned many of the things I now know. Thanks to Alina von Davier for always being receptive and willing to collaborate on projects. Thanks to Ernesto San Martín for always being there to discuss the most interesting problems in statistics and especially in psychometrics. I would also like to thank all of my students for working with me on their master's theses on equating. I would especially like to thank my parents, Luis and Iris, who have always given whatever is necessary for my well-being. Thanks to my brother, Carlos, for being my eternal best friend whom I can always trust. For all the time that I have stolen from you and for all of your understanding, thanks Fanny for being my partner for all these years. Lastly, for being the *non-equatable ones*, thanks to my children Javiera and Renato. This book is dedicated to both of you.

Marie's Acknowledgments. I would like to thank all of those whom I have worked with within the equating research field. I have learned many different things from all of you. A few persons should be mentioned. I am thankful to Wim van der Linden, who got me interested in the equating research field in 2006. I am grateful to Alina von Davier who introduced me to kernel equating and all of the new ideas we have worked with and will work with in the future. I am also thankful to Björn Andersson, who programmed the majority of **kequate** and continues to improve the package with his own research ideas. I would also like to thank my parents Britt and Ulf, for always being there for me. Finally, I am grateful to my family – my husband Jonas and my two daughters Emelie and Hanna – for their love and support and for being the best in the world.

Santiago, Chile
Umeå, Sweden
October 2016

Jorge González
Marie Wiberg

References

Albano, A. D. (2016). equate: An R package for observed-score linking and equating. *Journal of Statistical Software, 74*(8), 1–36.

Andersson, B., Bränberg, K., & Wiberg, M. (2013). Performing the kernel method of test equating with the package kequate. *Journal of Statistical Software, 55*(6), 1–25.

Battauz, M. (2015). equateIRT: An R package for IRT test equating. *Journal of Statistical Software, 68*(7), 1–22.

Choi, S., Gibbons, L., & Crane, P. (2011). lordif: An R package for detecting differential item functioning using iterative hybrid ordinal logistic regression/item response theory and monte carlo simulations. *Journal of Statistical Software, 39*(8), 1–30.

ETS (2011). KE-software (Version 3) [Computer software].

González, J. (2014). SNSequate: Standard and nonstandard statistical models and methods for test equating. *Journal of Statistical Software, 59*(7), 1–30.

Kolen, M., & Brennan, R. (2014). *Test equating, scaling, and linking: Methods and practices* (3rd ed.). New York: Springer.

Moses, T., & von Davier, A. (2011). A SAS IML macro for loglinear smoothing. *Applied Psychological Measurement, 35*(3), 250–251.

R Core Team (2016). *R: A language and environment for statistical computing*. Vienna, Austria: R Foundation for Statistical Computing.

von Davier, A. A., Holland, P., & Thayer, D. (2004). *The kernel method of test equating*. New York: Springer.

Weeks, J. P. (2010). plink: An R package for linking mixed-format tests using IRT-based methods. *Journal of Statistical Software, 35*(12), 1–33.

Contents

Acronyms

1PL	One-parameter logistic
2PL	Two-parameter logistic
3PL	Three-parameter logistic
AIC	Akaike's information criterion
BIC	Bayesian information criterion
BNP	Bayesian nonparametric
CB	Counterbalanced
CDF	Cumulative distribution function
CE	Chained equating
CINEG	Common item nonequivalent groups
CRAN	Comprehensive R Archive Network
CTT	Classical test theory
DBPP	Dependent Bernstein polynomial process
DF	Design function
DS	Double smoothing
DTM	Difference that matters
EG	Equivalent groups
FT	Freeman–Tukey residuals
HPD	Highest posterior density
ICC	Item characteristic curve
IRT	Item response theory
IRTTSE	IRT true-score equating
IRTOSE	IRT observed-score equating
MAD	Mean absolute deviation
MCMC	Markov chain Monte Carlo
MLE	Maximum likelihood estimator
MSD	Mean signed difference
MSE	Mean squared error
NEAT	Nonequivalent groups with anchor test
NEC	Nonequivalent groups with covariates
PDF	Probability density function

PRE	Percent relative error
PSE	Poststratification equating
RMSE	Root mean squared error
SE	Standard error
SEE	Standard error of equating
SEED	Standard error of equating difference
SG	Single group

List of Symbols

List of notation used in the book:

α	Person parameter in an IRT model
α	Sensibility parameter
α	Shape parameter in the beta distribution
θ	Ability
θ	Index for the true equating transformation in local equating
γ_P	Slope regression coefficient in the Tucker equating method and ratio used in nominal weights and Levine true-score equating on population P
γ_Q	Slope regression coefficient in the Tucker equating method and ratio used in nominal weights and Levine true-score equating on population Q
ω_j	Vector of item parameters for item j
Φ	Standard normal cumulative distribution function
ϕ	Standard normal density function
φ	The equating transformation
$\varphi_{\mathbf{z}_f,\mathbf{z}_t}$	Covariate-dependent equating transformation
Θ	Parameter space
ρ_j	Indicator variable in the penalty function
ℓ	Log-likelihood function
λ_j	Local bandwidth factor
λ	Parameter indexing the distribution of a discrete random variable
μ_X, σ_X	Mean and standard deviation of X
μ_Y, σ_Y	Mean and standard deviation of Y
μ_{XP}	Mean of X in population P
μ_{XQ}	Mean of X in population Q
μ_{YP}	Mean of Y in population P
μ_{YQ}	Mean of Y in population Q
μ_{XT}	Mean of X in target population T

μ_{YT}	Mean of Y in target population T
Ψ	Parameter indexing the law of stochastic processes
σ_{XT}	Standard deviation of X in target population T
σ_{YT}	Standard deviation of Y in target population T
$X(h_X)$	Continuized random variable
$\mathfrak{F}$	Function space
$\mathscr{M}(\theta)$	Model parametrized by θ
$\mathscr{X}$	Sample space for score X
$\mathscr{Y}$	Sample space for score Y
$\mathscr{Z}$	Covariate space
$\mathscr{A}$	Sample space for anchor score A
$\mathscr{C}$	Set of common items under the NEAT design
A	Anchor test
A, B	Equating coefficients in IRT parameter linking
a	Item discrimination parameter in IRT models
a	Anchor test score
A	Random variable for anchor test score
b	Item difficulty in IRT models
$\mathbf{b}$	Vector of constants used to specify a log-linear model for score probabilities
$\mathbf{B}$	Matrix of constants used to specify a log-linear model for score probabilities
c	Pseudo guessing parameter in IRT models
C	Covariates
D	Scaling constant
H	Distribution function
$h_{\mathbf{z}}$	A known [0, 1]-valued bijective continuous function defined on $\mathbb{R}$
$\mathscr{H}$	The set of $h_{\mathbf{z}}$
$E_P(X\|A)$	Conditional expectation of X given A in population P
$E_Q(X\|A)$	Conditional expectation of X given A in population Q
$E_P(Y\|A)$	Conditional expectation of Y given A in population P
$E_Q(Y\|A)$	Conditional expectation of Y given A in population Q
$f_{XT}(x)$	Synthetic target distribution for X
$f_{YT}(y)$	Synthetic target distribution for Y
$f_X(x\|a)$	Conditional distribution of X given anchor test score $A = a$
$f_Y(y\|a)$	Conditional distribution of Y given anchor test score $A = a$
$f_{AP}(a), f_{AQ}(a)$	Marginal distributions of A in population P and Q, respectively
F_{AP}	Anchor test score distribution in population P
F_{AQ}	Anchor test score distribution in population Q
F_{AT}	Anchor test score distribution in population T
$F_{X\|A}$	Conditional score distribution for test X given anchor test A
$F_{Y\|A}$	Conditional score distribution for test Y given anchor test A
F_X	Score distribution for test X
F_Y	Score distribution for test Y

$F_{\mathbf{z}}$	Score distribution indexed by covariates
$F_{h_X}(x)$	CDF of continuized random variable $X(h_X)$
$F_{h_Y}(y)$	CDF of continuized random variable $Y(h_Y)$
$\hat{f}_{h_X}(x_j)$	Estimated density function of continuized random variable $X(h_X)$
$\hat{f}_{h_Y}(y_j)$	Estimated density function of continuized random variable $Y(h_Y)$
$F_{XP}(x\|a)$	Conditional distribution of X in population P given anchor test score
$F_{XQ}(x\|a)$	Conditional distribution of X in population Q given anchor test score
$F_{YP}(y\|a)$	Conditional distribution of Y in population P given anchor test score
$F_{YQ}(y\|a)$	Conditional distribution of Y in population Q given anchor test score
F_{XT}	Distribution of X in target population T
F_{YT}	Distribution of Y in target population T
$p(\theta)$	Prior distribution
$p(\theta \mid x)$	Posterior distribution
$p(x \mid \theta)$	Likelihood
G_1, G_2	Group 1, Group 2
Hcrit	Haebara criterion
h_X, h_Y	Bandwidths in $X(h_X)$ and $Y(h_Y)$, respectively
$\mathbf{J}_{\varphi}$	Jacobian of the equating transformation
$\mathbf{J}_{DF}$	Jacobian of the design function
J	Total number of possible scores in test form X
J_x	Number of items in test form X
J_y	Number of items in test form Y
J_a	Number of items in test form A
K	Total number of possible scores in test form Y
K	Kernel chosen under the kernel equating framework
k	Constant in the penalty function in kernel equating
k	A discrete random variable
L	Total number of possible scores in test form A
L_r, L_s	Sum limits that serve to accommodate cross moments in polynomial log-linear models
n_j	Frequency of test score j
n_X, n_Y	Sample sizes of test takers for tests X and Y, respectively
$\mathbf{p}$	Vector of score probabilities
$\text{PEN}(h_X)$	Penalty function for h_X
$\Pr(X = x\|A = a)$	The probability of X given $A = a$
$\Pr(Y = y\|A = a)$	The probability of Y given $A = a$
r_j, s_k	Score probabilities for score X and Y
$\mathbf{r}, \mathbf{s}$	Vectors of score probabilities
$\mathbb{R}$	Set of real numbers

SLcrit	Stocking–Lord criterion
t	Failure time
t	Test score
T	Random variable denoting the failure time
T	Target population
T^{τ_X}, T^{τ_Y}	Test characteristic functions
τ_X, τ_Y, τ_A	True score for X, Y, and A, respectively
T_r, T_s	Highest degree of log-linear polynomial
V	A random variable
w_P, w_Q	In the NEAT design, weights for population P and Q in the synthetic population
X,Y	Test forms X and Y
X, Y	Random variable for test scores from X and Y
x_i	A possible test score value for X
y_i	A possible test score value for Y
$\mathbf{z}$	Covariate vector
π	Bernoulli probability parameter
π	Item characteristic function
π	Prior probability measure

Chapter 1
General Equating Theory Background

Abstract This chapter provides a general overview of equating and offers a conceptual and formal mathematical definition of equating. The roles of random variables, probability distributions, and parameters in the equating statistical problem are described. Different data collection designs are introduced, and an overview of some of the equating methods that will be described throughout the book is also presented.

1.1 Introduction

In different fields, it is usually of interest to compare measures coming from a common phenomena: "*The comparability of measurements made in differing circumstances by different methods and investigators is a fundamental pre-condition for all of science*" (Dorans and Holland 2000). Particularly in the field of educational measurement, the comparability of test scores is of interest because these scores are used to make important decisions in various settings. When making decisions, it is usually the case that individuals are compared on certain features that are measured as scores such that a score comparison implies a comparison among individuals. Scores are important in determining academic admissions, whether scholarships should be granted, to monitor progress in achievement, to determine competencies on a particular task, etc. Sometimes, score results do not lead to serious consequences, but when they are used for making decisions, it is important to report scores in a fair and precise way.

Mainly due to security reasons, it is common for measurement programs to produce different forms of a test that are intended to measure the same attribute. Although the test constructors try their best to produce test forms as parallel[1] as possible, differences in the difficulty of the forms are unavoidable. Thus, it is very possible that some test takers are administered an easier form of a test that is supposed to measure the same characteristic in all available test forms. When

[1]It is enough to consider test forms to be *parallel* if they measure the same attribute in a different way, for instance, by using different items. A more technical definition of parallel forms can be found in Lord (1964).

J. González, M. Wiberg, *Applying Test Equating Methods*,
Methodology of Educational Measurement and Assessment,
DOI 10.1007/978-3-319-51824-4_1

the magnitude of such characteristics is important for decisions, it is not fair to assign higher scores to some group of the population only because the test form administered to them was easier than another test form. In other words, if test forms were *equated* then it would be fair to give them to different people and treat the scores as if they come from the same test.

1.1.1 A Conceptual Description of Equating

Score differences are not exclusively due to differences in the difficulty of the tests forms. It might happen that a group of test takers is simply more able than the other in which case a comparison between groups by using test scores is confounded by the ability of individuals in the two groups. In this sense, a definition of *equating* should consider scores on two test forms as equivalent in a group of test takers if they represent the same relative position in the group (Livingston 2004).

The process of equating tests involves not only the adjustment of scores so that differences in difficulty can be compensated for, but it should also consider initial aspects so as to carefully construct the tests (i.e., test forms that differ in content or have different numbers of items cannot be considered parallel). In this book, the focus is on the first aspect. After disentangling the confounding between differences in group ability and differences in difficulty of tests forms, statistical models and methods will be used to adjust scores (rather than tests) and thus compensate for differences in the difficulty of the forms. We will thus define equating as a family of statistical models and methods that are used to make test scores comparable among two or more versions of a test, so that scores on these different test forms, which are intended to measure the same attribute, may be used interchangeably (see, e.g. Holland and Rubin 1982; von Davier et al. 2004; Dorans et al. 2007; von Davier 2011; Kolen and Brennan 2014).

1.1.2 A Statistical Model View of Equating

Our perspective is to view equating as statistical models because random variables, probability distributions and parameters are indeed involved in the process. We believe that the statistical theory viewpoint of equating (von Davier 2011; van der Linden 2011; González and von Davier 2013; Wiberg and González 2016) is useful not only for understanding the different methods that will be described in the subsequent chapters, but also because it opens new research possibilities on equating. With this approach, the score scales of two test forms that are to be equated will be seen as the sample spaces of two random variables representing the scores from both forms, respectively. The equating problem will thus be reduced to finding an appropriate function to map the scores from one scale into their equivalents in the scale of the other.

In the following section we introduce the general definition of a statistical model and discuss the basics concepts of parameters, parameter space, and sampling space. This is followed by a discussion about the role of random variables, probability distributions and parameters as involved in equating in this general framework.

1.2 Statistical Models

1.2.1 General Definition, Notation, and Examples

Statistical models assume that observed data are realizations of random variables following some probability distribution. Thus, statistical models are used to learn about the data generating mechanism. The random variables $X_1, \ldots, X_n$ are supposed to follow a distribution F_θ that is indexed by a *parameter* θ defined on a *parameter space* Θ. Let $x_1, \ldots, x_n \sim F_\theta$ be the observed data defined on a sample space $\mathscr{X}$. A statistical model can compactly be written as (e.g., Fischer 1922; Cox and Hinkley 1974; McCullagh 2002)

$$\mathscr{F} = \{\mathscr{X}, F_\theta : \theta \in \Theta\}, \tag{1.1}$$

where the family $\mathscr{F}$ is the collection of all probability distributions on $\mathscr{X}$ indexed by a parameter θ. The sample space $\mathscr{X}$ is the set of all possible observed data.

Example 1.1 (The linear regression model) Consider a simple linear regression model $y = \beta_0 + \beta_1 x + \varepsilon$ with $\varepsilon \sim N(0, \sigma^2)$. With this specification, the statistical model can be written as

$$\mathscr{F} = \{\mathbb{R}, N(\beta_0 + \beta_1 x, \sigma^2) : (\beta_0, \beta_1, \sigma^2) \in \mathbb{R}^2 \times \mathbb{R}^+\},$$

where $\mathbb{R}$ is the set of real numbers, $N(a, b)$ represents the normal distribution with mean a and variance b, and $\mathbb{R}^+$ is the set of positive real numbers (excluding zero). In this case, the probability model generating the data is the normal distribution, which is indexed by the parameters $\boldsymbol{\theta} = (\beta_0, \beta_1, \sigma^2)$. For known values of x, the parameter $\boldsymbol{\theta}$ fully characterizes the data generating mechanism which, in this case, produces the outcome variable y.

Example 1.2 (Fixed-effects Item Response Theory (IRT) models) Let X_{ij} be the random variable denoting the answer of test taker i to item j on a test. Fixed-effects IRT models are specified through the following two conditions: (i) mutual independence of X_{ij}, $i = 1, \ldots, n_x, j = 1, \ldots, J_x$, where $X_{ij} = 1$ when test taker i correctly answers item j , and $X_{ij} = 0$ otherwise; (ii) $X_{ij} \sim \text{Bernoulli}[\pi(\alpha_i, \omega_j)]$, where $\alpha_i \in \mathbb{R}$ corresponds to the person parameter, $\omega_j \in \Omega \subset \mathbb{R}^K$ corresponds to the item parameter, and π is a known and strictly increasing function in α_i for all

$\omega \in \Omega$. For instance, for the one parameter logistic (1PL) IRT model, $\omega_j = \beta_j$ and we have

$$\mathscr{F} = \{\{0,1\}, \text{Bernoulli}[\pi(\alpha_i, \beta_j)] : (\alpha_i, \beta_j) \in \mathbb{R} \times \mathbb{R}\},$$

where $\pi(\alpha_i, \beta_j) = \Pr(X_{ij} = 1 \mid \alpha_i, \beta_j) = \exp(\alpha_i - \beta_j)/1 + \exp(\alpha_i - \beta_j)$. In this case, the underlying probability model generating the data is the Bernoulli distribution indexed by $\boldsymbol{\theta} = (\alpha_i, \beta_j)$.

In the first example, the parameters of interest are directly related to the probability distribution that generates the data. In the second, person and item parameters are related to the Bernoulli distribution through π. In both cases, the data generating mechanism is fully characterized by $\boldsymbol{\theta}$.

Sometimes the parameters of interest are functions of other parameters that are involved in the model, as is shown in the following example.

Example 1.3 (Odds-ratio model) Consider a 2×2 contingency table with entries $X_{11}, X_{12}, X_{21}, X_{22}$. When the total number of counts n is known, the random variable $\mathbf{X} = (X_{11}, X_{12}, X_{21}, X_{22})$ conditional on n is multinomial. We write $\mathbf{X} \mid n \sim Mult(n; \boldsymbol{\theta})$ where $\boldsymbol{\theta} = (\pi_{11}, \pi_{12}, \pi_{21}, \pi_{22})$ are the cell probabilities. To test the raw-column independence, a commonly used parameter of interest is the odds ratio $OR = \frac{\pi_{11}\pi_{22}}{\pi_{12}\pi_{21}}$. Although the statistical model involves a family of multinomial distributions indexed by $\boldsymbol{\theta} \in [0,1]^4$, the parameter of interest is the OR that is not directly related to the data generating mechanisms but is instead a function of other parameters involved in the model.

Many other statistical models that are often used (e.g., factor analysis models, structural equation models, ANOVA models, log-linear models, etc) can be written in this compact way and can thus be shown to be a part of this general definition.

1.2.2 Types of Statistical Models

When the data generating mechanism (i.e., the probability model) is fully characterized by the parameter θ, then the model is said to be a *parametric* model. Formally, a parametric model is one where the dimension of the parameter space Θ is finite, meaning that $\theta \in \Theta \subset \mathbb{R}^d$ where d is a finite number. Sometimes the parameter characterizing the statistical model is a mixture of both finite and infinite-dimensional parameters (e.g., Schervish 1995, Section 1.6). In such a case, we denote the model as *semiparametric*. Yet another possibility is when no finite parameters characterize the probability distributions, but they are themselves the parameter of interest. The *nonparametric* approach starts by focusing on spaces of distribution functions, denoted by $\mathfrak{F}$, so that uncertainty is expressed on F itself (Tsiatis 2007).

To distinguish among these three types of models, the following notation is used to write the general definition for statistical models given in Eq. (1.1).

(i) *Parametric model*:

$$\mathscr{F} = (\mathscr{X}, F_\theta : \theta \in \Theta \subset \mathbb{R}^d)$$

(ii) *Semiparametric model*:

$$\mathscr{F} = (\mathscr{X}, F_\theta : \theta = (\theta_1, \theta_2), \theta_1 \in \Theta_1 \subset \mathbb{R}^d; \theta_2 \in \Theta_2 = \mathfrak{F})$$

(iii) *Nonparametric model*:

$$\mathscr{F} = (\mathscr{X}, F : F \in \mathfrak{F})$$

Note that in (i) and (ii) d is a finite number such that Θ and Θ_1 are finite-dimensional parameter spaces. In (iii), however, the parameter space Θ is typically the set of all probability distributions on $\mathscr{X}$ so that it constitutes a function space that we denote by $\mathfrak{F}$.

Example 1.4 (The linear regression model (cont'd)) Both semiparametric and nonparametric versions of the (normal) linear regression model can be defined. In the first case, the assumption of linearity on the mean of the conditional distribution can be relaxed, leaving the data to "speak for itself" about its functional form. In this case, we have $y = m(x) + \varepsilon$ where $m(\cdot)$ is a functional parameter that has to be estimated from the data. As before, $\varepsilon \sim N(0, \sigma^2)$ so σ^2 constitutes the parametric part of the model. Under this specification, the semiparametric statistical model becomes

$$\mathscr{F} = \{\mathbb{R}, N(m(x), \sigma^2) : (m(\cdot), \sigma^2) \in \mathfrak{F} \times \mathbb{R}^+\}.$$

Note that the parameters $\boldsymbol{\theta} = (m(\cdot), \sigma^2)$ completely characterize the data generating mechanism.

Example 1.5 (Survival analysis models) Let T be the random variable denoting the time at which an element fails. The main parameter of interest in this setting is the so-called *survival function*, which is defined as $S(t) = \Pr(T > t) = 1 - F_T(t)$. This functional parameter has the particularity that it is itself a function of another function: the cumulative distribution function (CDF) of the failure times $F(t)$. Having a sample of failure times $t_1, \ldots, t_n$, the main parameter of interest can be estimated by first estimating $F(t)$ and thus we have $\hat{S}(t) = 1 - \hat{F}(t)$. The data generating mechanism in this case is described by an infinite number of parameters.

Examples of semiparametric and nonparametric IRT models can be found in Duncan and MacEachern (2008), Miyazaki and Hoshino (2009), and San Martín et al. (2011). Example 1.5 shows that the parameters of interest that characterize the data generating mechanisms are not always finite elements $\boldsymbol{\theta}$ but instead the probability distributions themselves are of interest.

Note that no matter the type of model used, the aim of either *classical* or *Bayesian* statistical inference is to learn about the data generating mechanism that produced the data. The parameters of interest, which characterize the probability distributions in the models, have to be estimated from the data, which at the same time are supposed to be generated by the corresponding probability distribution. As we have seen, sometimes the probability distributions are themselves of interest and constitute the main parameter to be estimated from the data. Equating models are no exception to this general formulation as will be demonstrated in the next section.

1.2.3 Mathematical Statistics Formulation of the Equating Problem

As with any other statistical model, random variables, probability distributions, and parameters (either finite or infinite dimensional), also play a role in equating. In this section we introduce the notation and formal specification of these elements in the context of equating.

Although in many cases more than two test forms are to be equated, for ease of exposition we will consider only two test forms and denote them as X and Y. It will be assumed that these two forms are administrated to n_x and n_y randomly sampled test takers respectively. The scores obtained from test form X are denoted by X_i $(i = 1, \ldots, n_x)$, and those obtained from Y as Y_j $(j = 1, \ldots, n_y)$. The X and Y scores are assumed to be random variables that are accordingly defined on sample spaces $\mathscr{X}$ and $\mathscr{Y}$, and following the distributions F_X and F_Y, respectively. In this case, the sample spaces constitute the sets of all possible values for the scores in both test forms. For instance, if test form X is composed of 30 binary scored items and scores are assumed to be the total number of correct answers, then $\mathscr{X} = \{0, 1, 2, \ldots, 30\}$. Note however that other scoring rules such as the ones used in IRT models can produce different types of scores (i.e., continuous rather than discrete scores). Methods for equating such types of score scales will be discussed in Chap. 5. In what follows, the definitions will be given for what we call *observed scores*, which will always be the total number of correct answers on a test. The actually observed score data will be assumed to be realizations of the random variables X and Y following distributions $x_1, \ldots, x_{n_x} \sim F_X(x)$ and $y_1, \ldots, y_{n_y} \sim F_Y(y)$.

The statistical problem in equating consists in modeling the relationship between a score on one test form and its corresponding score in another form. Mathematically, this means that a function has to be defined that takes values in $\mathscr{X}$ and gives as a result a value on $\mathscr{Y}$. The following definition is general for any equating function transformation.

Definition 1.1 Let $\mathscr{X}$ and $\mathscr{Y}$ be two sample spaces. A function $\varphi : \mathscr{X} \mapsto \mathscr{Y}$ will be called an *equating transformation*.

The sample spaces can be of different dimensions, and this allows us to consider two test forms that differ in the number of items.[2] From Definition 1.1, it follows that the equating transformation maps the scores on the scale of one test form into the scale of the other. Note that for simplicity, Definition 1.1 assumes that an equating is made to map $\mathscr{X}$ on $\mathscr{Y}$, but the reverse equating is also possible thus imposing that φ is a function that produces a symmetric equating in the sense that if φ equates X to Y, then φ^{-1} equates Y to X. The equating transformation will constitute the object of primary interest in the equating problem.

In test theory, it is assumed that two parallel tests forms X and Y measure the same characteristic (ability) without knowing the mathematical relation between $\mathscr{X}$ (the X score scale) and $\mathscr{Y}$ (the Y score scale). As is seen from Definition 1.1, the equating transformation provides such a relationship. An easy first step to learn about the connection between $\mathscr{X}$ and $\mathscr{Y}$ would be to examine the score distributions that result from observed data that have been generated by test takers taking the tests. If φ equates X to Y, then the distribution of Y and the distribution of the transformed scores should be the same. Following these ideas, Braun and Holland (1982) gave a definition of equating that we reproduce in the next section. All methods that are described in this book follow this definition.

1.2.4 Mathematical Form of the Equating Transformation

Definition 1.1 establishes that the equating transformation maps the scores of one test form into the scale of another. However, an explicit mathematical form for φ has not been given yet. This form will depend on the definition of equating or, in other words, in the way we operationalize the idea of *treating scores as if they come from the same test.* Throughout this book we will use the definition given by Braun and Holland (1982) which we reproduce here using our notation.

Definition 1.2 (Braun and Holland's (1982) definition of equating) Let X and Y be two test forms both generating score data X and Y, respectively. Forms X and Y are said to be equated in population T by $\varphi(x)$ if $F_Y(y) = F_{\varphi(x)}(y)$.

Definition 1.2 indicates that two scores x and $y = \varphi(x)$ are equated if the distribution of Y and that of the transformed (equated) scores $\varphi(X)$ are the same. Under this definition, we can obtain an explicit form for φ that equates X to Y in T. As a matter of fact, if x and y are the quantiles in the distributions of tests scores X and Y for an arbitrary common cumulative proportion p of the population, i.e., $x(p) = F_X^{-1}(p)$

[2]In practice such difference should be small (Braun and Holland 1982, p. 16).

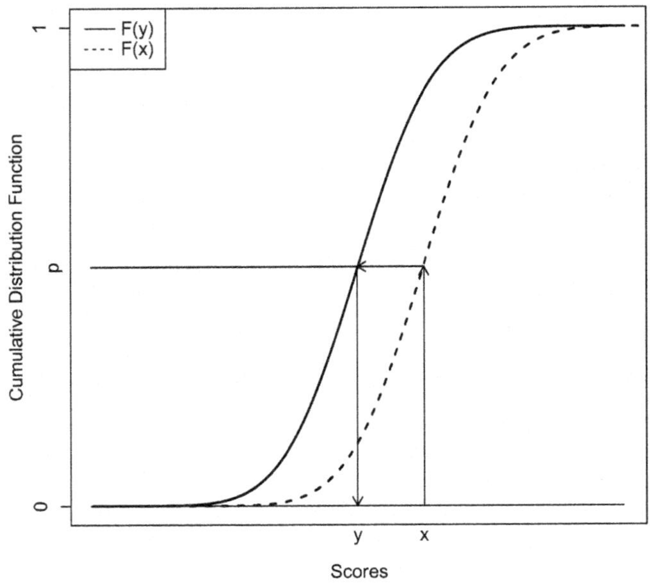

Fig. 1.1 Graphical representation of the equipercentile equating transformation. A score x in test form X is mapped into a score on the scale of test form Y using $y = \varphi(x) = F_Y^{-1}(F_X(x))$

and $y(p) = F_Y^{-1}(p)$, it follows that, for every $p \in [0, 1]$, with the requirement that $F_Y(y) = F_{\varphi(x)}(y)$, an equivalent score y on test Y for a score x on X can be obtained as

$$y = \varphi(x) = F_Y^{-1}(F_X(x)). \tag{1.2}$$

Although Eq. (1.2) can be used in general for the comparison of any two samples or distributions of random variables (see, e.g., Wilk and Gnanadesikan 1968), in the equating literature the function φ is known as the equipercentile transformation (Angoff 1971). A graphical representation of the equipercentile equating transformation is shown in Fig. 1.1. The heuristic behind this definition is simple: to define equated scores as those that have the same percentile rank in a given group of test takers.

1.2.5 Continuization

All of the equating methods that will be described in subsequent chapters are based on the mathematical form shown in Eq. (1.2) and thus suffer from the following problem. For multiple choice tests, where answers are coded as either correct or incorrect, one of the most commonly used test scores in measurement programs are the so-called sum score, which corresponds to the total number of correct answers on the test. Because the possible values that sum scores can take are consecutive integers, an evident problem with Eq. (1.2) is the discreteness of the scores' distributions, rendering their corresponding inverses unavailable. Thus, in common practice estimates of φ are based on continuous approximations of the originally discrete distributions F_X and F_Y. In the equating literature, this practice is called *continuization*, and it requires one to actually "continuize" the discrete score distributions F_X and F_Y in order to properly use Eq. (1.2) for equating. Typical continuization methods used in equating include linear interpolations, kernel smoothing techniques, and continuized log-linear methods.

1.2.6 Requirements for Comparability of Scores

So far, $\varphi(x)$ functions as a mathematical rule that takes elements in $\mathcal{X}$ and return their equivalents in $\mathcal{Y}$. For test scores to be considered interchangeable, it is obvious that the elements in $\mathcal{X}$ and $\mathcal{Y}$ must be at least *similar* in some sense. The extent of such similarity is a discussed topic in the research field of equating. The following requirements are needed for the mapping in Definition 1.1 to be validly called an equating: (i) *same construct*: the test forms being equated should measure the same construct; (ii) *reliability*: the test forms should be equally reliable; (iii) *symmetry*: the equating transformation to map $\mathcal{Y}$ into $\mathcal{X}$ should be φ^{-1} (i.e., the inverse of the equating transformation defined in Definition 1.1 to map $\mathcal{X}$ into $\mathcal{Y}$); (iv) *equity*: if X and Y have been equated, then the administered form should not be a concern for test takers; and (v) *group invariance*: the equating function φ should be invariant if score data from different groups in the population are used for estimation. These requirements have been listed in completeness, for instance, in Kolen and Brennan (2014), von Davier et al. (2004), Dorans and Holland (2000), and some of them have origins from Angoff (1971), Lord (1980), and Petersen et al. (1989).

1.2.7 Assessing the Uncertainty of Equating Results

Because random samples of individuals produce realizations of random score variables, uncertainty in the estimation of the equating transformation appears naturally. Although from the general setup outlined so far in this chapter it might appear evident that the main object of inference is the equating transformation, an obtained equating has typically been evaluated with different measures within a particular framework. A well accepted index used to measure the uncertainty in the estimation of φ is the *standard error of equating* (SEE). The SEE corresponds to the standard deviation of the equated scores over a number of hypothetical replications of an equating. As it will be seen in subsequent chapters, the way in which the SEE is calculated when using different equating methods varies. Note, however, that the viewpoint of the authors is that the equating transformation is the parameter of interest to conduct statistical inference in the equating problem, and estimations of the equating transformation need to be statistically assessed as with any standard evaluation of an estimator of an unknown parameter in statistics (Wiberg and González 2016). In Chap. 7 we explicitly show how it is possible to define a "true" equating transformation that can be evaluated against the estimated equating transformation using measures such as bias, mean square error (MSE) and standard error (SE), as is customary in ordinary statistical inference.

1.3 Collecting Data in Equating

In practice, score data are needed for the estimation of φ. Because differences in the distribution of scores can be attributed either to differences in difficulty of the test forms administered or due to differences in ability in the test takers, it is necessary to disentangle the confusion before any linking between $\mathscr{X}$ and $\mathscr{Y}$ takes place. As mentioned above, equating will correct for differences in difficulty of the tests forms, so we first need a way to correct for differences in ability. The way such correction is made will be guided by the way the score data are collected.

Suppose we have many test takers each with both x and y score data. Because the same group of people have both test scores, differences in ability can be assumed to be controlled for and thus by using the pairs (x, y), we can find $y = \varphi(x)$ approximately. If it is not possible to obtain x and y from the same test takers, we can assume that test takers with test score x are a random sample from the same population as test takers with test score y. In this way, we are assuming that the groups are *equivalent*. If the test forms are administered to different groups of test takers that cannot be considered to be equivalent, differences in ability between the groups can arise. In this case, the common solution is to use the scores of a set of common items, typically referred to as an *anchor test* (or any variable that can account for differences in abilities) whose scores are used to measure and control for these differences. An anchor test should reflect the content representation and the difficulty level of the whole test. It is widely accepted that an anchor test used in test equating benefits from being a miniature version of the current test, and thus it is sometimes called a *minitest* (e.g., Kolen and Brennan 2014), although some criticisms of this belief have been brought up by some authors (e.g., Sinharay et al. 2012).

In summary, because the differences in the distribution of scores can be attributed either to differences in the difficulty of the forms administered or to differences in ability in the test takers, it is necessary to control for ability differences before any equating take place. Depending on the situation, either common test takers or common items will be used to disentangle the confusion between these two factors.

The following sections discuss various equating designs for collecting score data that are commonly described in the equating literature.

1.3.1 Data Collection Designs in Equating

The observed score data can be thought of as a realization of a random variable that represents the score of a test taker belonging to a certain group in a certain population. Although in principle it is possible to obtain data from any number of different groups of test takers within any number of different populations, the exposition that follows will consider only two populations and two groups within each population. Likewise, only two test forms and one anchor test will be considered although in theory any number of them could be used for equating.

In defining the equating designs, the population of interest or *target* population from where score data are to be obtained will be denoted as T. If test takers cannot be considered to be equivalent but instead are considered to come from different populations, then the two different populations will be denoted as P and Q. In such a case, T will be some kind of mixture of both P and Q. Different equating data collection designs are described in the following sections.

1.3.1.1 Single Group Design

In the single group (SG) design, a unique sample group (G) of test takers ($n_x = n_y = n$) from a common population T is administered both test X and test Y. Because every test taker is administered both X and Y, the resulting data is a bivariate vector (x_i, y_i), $i = 1, \ldots, n$. One disadvantage of this design is the possible effects of fatigue or familiarity that depend on whether one test or another is administered first. Another issue is that the total test time can become long because each test taker is administered both X and Y.

1.3.1.2 Equivalent Groups Design

In the equivalent groups (EG) design, two independent groups of test takers (G_1 and G_2) are sampled from a common population T. Each group is administered only one of two test forms. Thus, the resulting data are two independent variables x_i ($i = 1, \ldots, n_x$) and y_j ($j = 1, \ldots, n_y$). When tests are given to groups at different time points, and the composition of test takers in the groups have changed over time, it has been shown that adopting an EG design is sometimes problematic because the assumption of equivalence might not hold (Lyrén and Hambleton 2011).

1.3.1.3 Counterbalanced Design

Similar to the EG design, in the counterbalanced (CB) design two independent groups of test takers are sampled from the same population T. The difference between the EG and the CB design is that in the CB design both groups are administered both test forms, but in different orders. Assuming that n_{G_1} and n_{G_2}

test takers are sampled in groups G_1 and G_2, respectively, the obtained data are two independent bivariate vectors (x_{1i}, y_{2i}) $(i = 1, \ldots, n_{G_1})$ and (x_{2j}, y_{1j}) $(j = 1, \ldots, n_{G_2})$. The subindexes 1 and 2 denote the order in which the tests were administered. In this case group 1 was first administered form X and then form Y, whereas group 2 was first administered form Y and then form X. Because it is possible to view this design as containing two EG designs on the one hand and two SG designs on the other, there are several possibilities to use the data from a CB design for equating. For a discussion on different possibilities of using these types of data, see von Davier et al. (2004). Although the order effect can be removed using this design, the time needed to administer two forms can be a problem.

1.3.1.4 Non Equivalent Groups with Anchor Test Design

In the non equivalent groups with anchor test (NEAT) design, two groups are independently sampled from different populations P and Q. Each group is administered either of the test forms X and Y, and a common anchor test form A is administered to both groups. Let a be the observed anchor test score, then the obtained data are two independent vectors (x_i, a_i) $(i = 1, \ldots, n_x)$ and (y_j, a_j) $(j = 1, \ldots, n_y)$. When anchor scores count for the total score reported, they are referred to as *internal*, whereas when they are used only for equating purposes and do not count for the reported score, they are referred to as *external*. This design is also referred to as common item nonequivalent group design (CINEG) by some authors (e.g., Kolen and Brennan 2014). A potential threat to the NEAT design is the fact that test takers might recognize anchor items and not take such items as seriously as the rest of the test. It should also be noted that not all measurement programs using standardized tests have anchor tests, and this makes the NEAT design impossible to use in some situations.

1.3.1.5 Non Equivalent Groups with Covariates Design

The non equivalent groups with covariates (NEC) design is an important alternative to the NEAT design in the case when no anchor test are available for equating (Wiberg and Bränberg 2015). As it is the case in the NEAT design, two groups are independently sampled from different populations P and Q and each is administered either of the test forms X and Y. In the absence of an anchor test form, the NEC design uses relevant covariates denoted by C that can account for differences in the groups of test takers. This design and the type of data obtained from it will be discussed in more detail in Chap. 4. A similar approach that also uses background information to form pseudo-equivalent groups has been proposed by Haberman (2015). New methods for the NEC design have recently been proposed by Sansivieri and Wiberg (2017) and Wallin and Wiberg (2017).

Table 1.1 shows a summary of the equating designs described above. Other similar schemas explaining some of these designs are encountered in von Davier

Table 1.1 Description of the equating designs with respect to populations, groups, and what information is gathered

		T				*P*				*Q*			
Design	Group	X	Y	A	C	X	Y	A	C	X	Y	A	C
SG	G	x	x										
EG	G_1	x											
	G_2		x										
CB	G_1	x	x										
	G_2	x	x										
NEAT	G_1					x		x					
	G_2										x	x	
NEC	G_1					x			x				
	G_2										x		x

Note: C = covariates, A = Anchor test, T = Target population, X = Test form X, Y = Test form Y, P = Population P, Q = Population Q, SG = single group, EG = equivalent groups, CB = counterbalanced, $NEAT$ = nonequivalent groups with anchor test, NEC = nonequivalent groups with covariates

et al. (2004, Chapter 2) and Kolen and Brennan (2014, Section 1.4). The score data obtained under these different designs will be used to estimate the equating transformation φ.

1.4 Some Examples of Equating Transformations

Different continuization methods define different statistical inference approaches that produce either parametric, semi-parametric, or nonparametric estimators of φ (González and von Davier 2013). In the following sections we give some examples of these estimators and leave the details for the subsequent chapters.

1.4.1 The Equipercentile Equating Function

The equipercentile equating transformation is formally defined exactly as in Definition 1.2, i.e.

$$\varphi(x) = F_Y^{-1}(F_X(x)). \tag{1.3}$$

Having observed score data $x_1, \ldots, x_{n_x} \sim F_X$ and $y_1, \ldots, y_{n_y} \sim F_Y$, a natural estimator for the CDFs involved in the calculation of φ is the nonparametric empirical distribution function. In fact, because we do not specify any particular parametric family of score distributions, the data generating mechanism is fully described by the two CDFs F_X and F_Y thus resulting in an equating estimator that is nonparametric by nature. Note, however, that in using this estimator, the problem of discreteness of the distributions persists and that is why linear interpolation has been used to continuize the obtained discrete distributions.

1.4.2 The Linear Equating Function

Various models that will be described in Chap. 3 utilize a linear equating function, including the mean, linear, Tucker and Levine equating (Kolen and Brennan 2014). The equating transformations are still based on the general definition given in Eq. (1.2). As a matter of fact, if a parametric location-scale family of score distributions is assumed for X and Y with means (μ_X, μ_Y) and standard deviations (σ_X, σ_Y) such that

$$F_X(x) = H\left(\frac{x - \mu_X}{\sigma_X}\right)$$

$$F_Y(y) = H\left(\frac{y - \mu_Y}{\sigma_Y}\right),$$

where H is some distribution function (e.g., Braun and Holland 1982; von Davier et al. 2004), then

$$\varphi(x; \boldsymbol{\theta}) = F_Y^{-1}(F_X(x; \mu_X, \sigma_X); \mu_Y, \sigma_Y) = \mu_Y + \frac{\sigma_Y}{\sigma_X}(x - \mu_X), \qquad (1.4)$$

In this case, the parameter $\boldsymbol{\theta} = (\theta_x, \theta_y) = (\mu_X, \mu_Y, \sigma_X, \sigma_Y) \in \Theta = \mathbb{R}\times\mathbb{R}\times\mathbb{R}^+\times\mathbb{R}^+$ characterizes the score distributions that generate the data leading to an equating estimator that is completely parametric. The method of moments leads to sample means and variances that are directly estimated from the observed data. In Chap. 3 we give a detailed account of these methods and illustrate their use in different equating designs.

1.4.3 The Kernel Equating Function

Kernel equating (von Davier et al. 2004) can be considered to be an example of a semi-parametric equating estimator in that both finite parameters and distribution functions are involved in the estimation of φ. The parametric part consists of score probability parameters $r_j = \Pr(X = x_j)$, and $s_k = \Pr(Y = y_k)$ with x_j and y_k taking values in $\mathcal{X}$, and $\mathcal{Y}$, respectively, and these are often considered to be the parameters of a multinomial distribution (see Sect. 2.3). On the other hand, the score distribution functions are estimated using nonparametric kernel smoothing techniques. The equating transformation used in kernel equating is written as

$$\varphi(x;\boldsymbol{\theta}) = F_{h_Y}^{-1}(F_{h_X}(x;\mathbf{r}),\mathbf{s}).$$

where **r** and **s** are the vectors with elements r_j and s_k, and h_X and h_Y are bandwidth parameters controlling the amount of smoothness. Kernel equating will be discussed in detail in Chap. 4. Other examples of semiparametric equating estimators that will be discussed in Chap. 6 are local equating (van der Linden 2011), and combinations of kernel and local equating (Wiberg et al. 2014).

1.5 R Packages That Are Used in This Book

A unique feature of this book is that it provides **R** code for performing all of the described equating methods. Four of the packages that will be used in the book can be considered "specialized" in the sense that they are particularly developed for equating. The **R** package **equate** (Albano 2016) implements traditional equating methods as well as smoothing of test score distributions. It will be used mainly in Chaps. 2 and 3. The **SNSequate** package (González 2014), implements both traditional and kernel equating methods and will be used in Chaps. 2, 3, 4, 5, 6 and 7. The **kequate** package (Andersson et al. 2013) is focused on kernel equating including IRT kernel equating and it will be used in Chaps. 2, 4, 6, and 7. We will also use **equateIRT** (Battauz 2015) in Chap. 5 when item parameter linking is described.

The other packages are not particularly focused on equating, but the models and methods implemented in them will be helpful in particular stages of some of the equating methods. Among the packages in this list, **ltm** (Rizopoulos 2006) and **mirt** (Chalmers 2012) will be used to estimate IRT models that will play an important role in some equating methods described in Chaps. 5, 6 and 7.

Finally, some of the methods that will be illustrated in the book are not yet part of any **R** package distribution. In this case, we will provide the raw **R** code that implements such method.

1.6 Summary and Overview of the Book

In this chapter we have outlined the main ideas behind equating using a formal mathematical statistics approach. Random variables, sample spaces, probability distributions, and parameters have been shown to indeed be part of the equating problem. Although this book is primarily oriented on practically applying equating, we believe that a formal definition of the elements and steps involved in equating will help to better understand the actual implementation of the methods. Through the remaining chapters of the book, the **R** software (R Core Team 2016) will be used to illustrate how the described equating methods can be implemented in practice. In Chap. 2 we describe the different data sets that will be used for illustrations. The way in which the score distributions are prepared to feed the **R** functions is also described in this chapter. Chapter 3 introduces traditional methods of equating including equipercentile, mean, and linear equating, and the **R** package **equate** (Albano 2016) is used to illustrate these methods. In Chap. 4 the kernel method of equating is described. The **R** packages **kequate** (Andersson et al. 2013) and **SNSequate** (González 2014) will be used to illustrate the kernel method of equating. Chapter 5 describes IRT equating methods, including IRT true-score and IRT observed-score equating. Methods that transform IRT scales rather than observed score scales are also described in this chapter. The **R** packages **SNSequate**, **equateIRT**, **mirt**, and **ltm** are used for illustrations. In Chap. 6 local equating is discussed and some relationships with other equating methods are described and exemplified using **SNSequate** and **kequate**.

Viewing equating as statistical models and the equating transformation as a functional parameter that has to be estimated from score data, enhances the possibilities in equating. Parametric, nonparametric, and semiparametric types of equating transformations can be introduced, and all types of inferences both from the classical (frequentist) and Bayesian point of view can indeed be conducted (González et al. 2015a,b). Some of these new possibilities are described in Chap. 7, and **kequate** and **SNSequate** are used for illustrations. The new developments in equating that are introduced include Bayesian nonparametric estimation of equating transformations, different bandwidth selection methods, and the use of alternative kernels and IRT in the kernel equating framework.

References

Albano, A. D. (2016). equate: An R package for observed-score linking and equating. *Journal of Statistical Software, 74*(8), 1–36.

Andersson, B., Bränberg, K., & Wiberg, M. (2013). Performing the kernel method of test equating with the package kequate. *Journal of Statistical Software, 55*(6), 1–25.

Angoff, W. H. (1971). Scales, norms and equivalent scores. In R. L. Thorndike (Ed.), *Educational measurement* (2nd ed., pp. 508–600). Washington, DC: American Council on Education. (Reprinted as Angoff WH (1984). *Scales, Norms and Equivalent Scores*. Princeton, NJ: Educational Testing Service.).

Battauz, M. (2015). equateIRT: an R package for IRT test equating. *Journal of Statistical Software, 68*(7), 1–22.

Braun, H., & Holland, P. (1982). Observed-score test equating: a mathematical analysis of some ETS equating procedures. In P. Holland & D. Rubin (Eds.), *Test equating* (Vol. 1, pp. 9–49). New York: Academic Press.

Chalmers, R. P. (2012). mirt: A multidimensional item response theory package for the R environment. *Journal of Statistical Software, 48*(6), 1–29.

Cox, D. R., & Hinkley, D. V. (1974). *Theoretical statistics*. London, UK: Chapman and Hall.

Dorans, N., & Holland, P. (2000). Population invariance and the equatability of tests: Basic theory and the linear case. *Journal of Educational Measurement, 37*(4), 281–306.

Dorans, N. J., Pommerich, M., & Holland, P. W. (2007). *Linking and aligning scores and scales*. New York: Springer.

Duncan, K., & MacEachern, S. (2008). Nonparametric Bayesian modelling for item response. *Statistical Modelling, 8*(1), 41–66.

Fischer, R. A. (1922). On the mathematical foundations of theoretical statistics. *Philosophical Transactions of the Royal Society of London, Series A, 222*, 309–368.

González, J. (2014). SNSequate: Standard and nonstandard statistical models and methods for test equating. *Journal of Statistical Software, 59*(7), 1–30.

González, J., Barrientos, A. F., & Quintana, F. A. (2015a). A dependent Bayesian nonparametric model for test equating. In R. Millsap, D. Bolt, L. van der Ark, & W.-C. Wang (Eds.), *Quantitative psychology research* (pp. 213–226). Cham: Springer International Publishing.

González, J., Barrientos, A. F., & Quintana, F. A. (2015b). Bayesian nonparametric estimation of test equating functions with covariates. *Computational Statistics & Data Analysis, 89*, 222–244.

González, J., & von Davier, M. (2013). Statistical models and inference for the true equating transformation in the context of local equating. *Journal of Educational Measurement, 50*(3), 315–320.

Haberman, S. J. (2015). Pseudo-equivalent groups and linking. *Journal of Educational and Behavioral Statistics, 40*(3), 254–273.

Holland, P., & Rubin, D. (1982). *Test equating*. New York: Academic Press.

Kolen, M., & Brennan, R. (2014). *Test equating, scaling, and linking: Methods and practices* (3rd ed.). New York: Springer.

Livingston, S. A. (2004). *Equating test scores (without IRT)*. Princeton, NJ: ETS.

Lord, F. (1964). Nominally and rigorously parallel test forms. *Psychometrika, 29*(4), 335–345.

Lord, F. (1980). *Applications of item response theory to practical testing problems*. Hillsdale, NJ: Lawrence Erlbaum Associates.

Lyrén, P.-E., & Hambleton, R. K. (2011). Consequences of violated equating assumptions under the equivalent groups design. *International Journal of Testing, 11*(4), 308–323.

McCullagh, P. (2002). What is a statistical model? (with discussion). *The Annals of Statistics, 30*, 1225–1310.

Miyazaki, K., & Hoshino, T. (2009). A Bayesian semiparametric item response model with Dirichlet process priors. *Psychometrika, 74*(3), 375–393.

Petersen, N. S., Kolen, M. J., & Hoover, H. D. (1989). Scaling, norming, and equating. *Educational Measurement, 3*, 221–262.

R Core Team (2016). *R: A language and environment for statistical computing*. Vienna, Austria: R Foundation for Statistical Computing.

Rizopoulos, D. (2006). ltm: An R package for latent variable modeling and item response theory analyses. *Journal of Statistical Software, 17*(5), 1–25.

San Martín, E., Jara, A., Rolin, J.-M., & Mouchart, M. (2011). On the Bayesian nonparametric generalization of IRT-type models. *Psychometrika, 76*(3), 385–409.

Sansivieri, V., & Wiberg, M. (2017). IRT observed-score with the non-equivalent groups with covariates design. In L. A. van der Ark, M. Wiberg, S. A. Culpepper, J. A. Douglas, & W.-C. Wang (Eds.), *Quantitative psychology – 81st annual meeting of the psychometric society, Asheville, North Carolina, 2016*. New York: Springer.

Schervish, M. J. (1995). *Theory of statistics*. New York: Springer.

Sinharay, S., Haberman, S., Holland, P., & Lewis, C. (2012). A note on the choice of an anchor test in equating. *ETS Research Report Series, 2012*(2), i–9.

Tsiatis, A. (2007). *Semiparametric theory and missing data*. New York: Springer.

van der Linden, W. J. (2011). Local observed-score equating. In A. von Davier (Ed.), *Statistical models for test equating, scaling, and linking* (pp. 201–223). New York: Springer.

von Davier, A. (2011). *Statistical models for test equating, scaling, and linking*. New York: Springer.

von Davier, A. A., Holland, P., & Thayer, D. (2004). *The kernel method of test equating*. New York: Springer.

Wallin, G., & Wiberg, M. (2017). Non-equivalent groups with covariates design using propensity scores for kernel equating. In L. A. van der Ark, M. Wiberg, S. A. Culpepper, J. A. Douglas, & W.-C. Wang (Eds.), *Quantitative psychology – 81st annual meeting of the psychometric society, Asheville, North Carolina, 2016*. New York: Springer.

Wiberg, M., & Bränberg, K. (2015). Kernel equating under the non-equivalent groups with covariates design. *Applied Psychological Measurement, 39*(5), 349–361.

Wiberg, M., & González, J. (2016). Statistical assessment of estimated transformations in observed-score equating. *Journal of Educational Measurement, 53*(1), 106–125.

Wiberg, M., van der Linden, W. J., & von Davier, A. A. (2014). Local observed-score kernel equating. *Journal of Educational Measurement, 51*, 57–74.

Wilk, M., & Gnanadesikan, R. (1968). Probability plotting methods for the analysis of data. *Biometrika, 55*(1), 1–17.

Chapter 2
Preparing Score Distributions

Abstract Depending on the equating data collection design, score data will be in the form of either univariate or bivariate distributions. In this chapter, we describe how to prepare the score distributions in order to read them into the different **R** packages that will be used. Presmoothing the score distributions as a first step in equating is also discussed. Known data sets appearing in the equating literature, as well as real data examples from an admissions test and an achievement test are described. The illustrations will use the three **R** packages **equate** (Albano, J Stat Softw 74(8):1–36, 2016), **kequate** (Andersson et al., J Stat Softw 55(6):1–25, 2013) and **SNSequate** (González, J Stat Softw 59(7):1–30, 2014).

2.1 Data

Throughout the book, different equating methods will be exemplified using available real data sets. Published data are used to reproduce various equating results appearing in the literature, and two new real data sets from an admissions test and an achievement test are used for illustrative purposes. All of these data sets and how they can be obtained are briefly described in the following sections.

2.1.1 Data from Kolen and Brennan (2014)

Kolen and Brennan (2014) describe and use two data sets to exemplify various methods of equating. The first data set consists of tests forms of the original ACT Mathematics test. The test contains 40 multiple-choice items scored incorrect (0) or correct (1). Test form X was administered to 4,329 examinees and test form Y to 4,152 examinees. This data set is presented in Table 2.5 of Kolen and Brennan (2014) as score frequencies. We will refer to these data as the `ACT` data set. Both the **equate** and **SNSequate** packages include the `ACT` data set as the objects `ACTmath` and `ACTmKB`, respectively. Both objects contain the score scale, score frequencies for test form X, and the score frequencies for test form Y.

J. González, M. Wiberg, *Applying Test Equating Methods*,
Methodology of Educational Measurement and Assessment,
DOI 10.1007/978-3-319-51824-4_2

The second data set, which we call `KB36`, consists of two 36-items test forms. Form X was administered to 1,655 examinees and form Y was administered to 1,638 examinees. Also, 12 out of the 36 items are common between both test forms (items 3, 6, 9, 12, 15, 18, 21, 24, 27, 30, 33, and 36). The `KBneat` object in the **equate** package contains the data. The same data, but in a more complete way is provided in the `KB36` object in the **SNSequate** package. The `KB36` object contains not only the score frequencies, but also the response patterns and item parameters estimates obtained from a 3PL (three parameter logistic) IRT model fitted to the two 36-items tests forms. `KB36` is a list with four elements containing binary data matrices of responses (`KBformX` and `KBformY`) and the corresponding parameter estimates that result from a 3PL IRT model fit to both data matrices (`KBformX_par` and `KBformY_par`). The 0-1 score data come with the distribution of the **CIPE** software (Kolen and Brennan 2014) that is freely available at https://www.education.uiowa.edu/centers/casma/computer-programs. The list of item parameters estimates can be found in Table 6.5 of Kolen and Brennan (2014).

2.1.2 Data from von Davier et al. (2004)

Other real data sets used in the literature are described in von Davier et al. (2004) for the EG, SG, CB and NEAT designs. For the EG design, data for two parallel 20-items mathematics tests given to two samples from a national population of test takers are given in Table 7.1 in von Davier et al. (2004). The `Math20EG` object in **SNSequate** contains raw sample frequencies of number correct scores for the two test forms. A similar test is used to illustrate the SG design with data appearing in Table 8.2 in their book that shows the corresponding bivariate score frequency distributions. These data are available as the `Math20SG` object in **SNSequate**. For the CB design, a data set from a small field study from an international testing program is used. This data set contains the observed scores for test form X (with 75 items) and Y (with 76 items) administered to two independent, random samples of test takers from a single population. These data are available in Chap. 9 in von Davier et al. (2004) and make up the `CBdata` object in **SNSequate**.

2.1.3 The `ADM` Admissions Test Data

The `ADM` data set contains information about an admissions test that is given twice a year and is used as an entrance test to universities and colleges. The test is composed of a verbal and a quantitative section, each having 80 multiple-choice items that are binary scored and equated separately. An anchor test of 40 items is given to a smaller sample of test takers in each administration in order to perform the equating for each section. To illustrate the methods under various equating data collection designs, different samples of two consecutive administrations are used. These samples have been stored as **R** data sets[1] that are available on this book's webpage. For the EG design, the `ADM1` and `ADM2` data sets both contain samples of 10,000 test takers. For the NEAT design, the `ADMneatX` data contains one sample from population P with 2,000 test takers, and the `ADMneatY` data contain one sample from population Q with 2,000 test takers who took the anchor test. For the SG design, a sample of 8,000 test takers who were administrated both test forms is available as the `ADM12` data set.

2.1.4 *The* `SEPA` *Test Data*

The `SEPA` data come from a private national evaluation system in Chile called SEPA (Sistema de Evaluación de Progreso del Aprendizaje; System of Assessment Progress in Achievement) administered by the measurement center MIDE UC. SEPA consists of tests specifically designed to assess achievement in students from first to eleventh grade in the subjects of Language and Mathematics. For each of the two subjects, there are two test forms within a particular grade in a year. The total number of items in the tests depends on the grade assessed, and goes from 25 (first grade) to 50 (eleventh grade). In each application, additional information such as test taker's gender, and school type (e.g., municipal, subsidized, private) is also available in the data base. The `SEPA` object in **SNSequate** contains the test scores in mathematics for 1,458 and 2,619 eight grade test takers administered test forms X and Y, respectively, where each test form contains 50 items.

2.2 Preparing the Score Data

As seen in Sect. 1.3.1 the type of score data to be used for equating will depend on the adopted equating data collection design. For instance, under the EG design, the observed score data are assumed to be the realizations of two independent random variables X and Y with x_i $(i = 1, \ldots, n_x)$ and y_j $(j = 1, \ldots, n_y)$, whereas under the SG design, the score data are in the form of a bivariate vector (x_i, y_i) $(i = 1, \ldots, n)$. Both the CB and NEAT design also produce bivariate score data. After tests have been administered, score data are stored in different ways. Some test score files contain the sum score of the test takers, other files contain the number of test takers at each test score (score frequencies) and yet others contain only the response patterns of test takers for each test form. Most of the **R** packages that are used for

[1]The data can also be obtained in different file formats from this book's webpage. Appendix A provides examples on how to read data files of different formats in **R**.

equating, analyze score distributions primarily as frequency table objects. In what follows, we first introduce the functions from the **equate** and **kequate** packages that are needed to create score frequency distributions. Examples are presented for the EG, SG and NEAT designs.

2.2.1 Functions to Create Score Frequency Distributions

The **equate** package uses score distributions structured as frequency tables in **R** within the class `freqtab`. The `freqtab()` and `as.freqtab()` functions are used to produce either univariate or bivariate score frequency distributions as table arrays using a vector or data frame of observed scores. A typical call to `freqtab()` reads as

```
freqtab(x, scales, design, ...)
```

where the argument `x` is either a vector of scores or a matrix whose raw entries correspond to the response patterns of each test taker. The argument `scales` is either a single sequence of numbers indicating the score scale for the test (i.e., for the case of univariate distributions) or a list containing two or more sequences of numbers indicating the multiple scales (i.e., under the SG, CB, or NEAT design). The assumed equating design is set using the argument `design` with default values `design = "eg"` (EG design) and `design = "ng"` (NEAT design) for univariate and multivariate score distributions, respectively. The SG and CB designs are specified using the options `"sg"` and `"cb"`, respectively.

The **kequate** package has the `kefreq()` function whose functionality is very similar to that of `freqtab()` in **equate**. A typical call to `kefreq()` reads as

```
kefreq(in1, xscores, in2, ascores)
```

where `in1` and `in2` are vectors containing scores on test form X and on test forms Y or A (the anchor test), respectively. The arguments `xscores` and `ascores` are used to indicate the score scales for tests X and Y or A, respectively. How to use the `kefreq()` function is illustrated in this chapter for the EG, SG and NEAT designs. An illustration of the NEC design is postponed until Chap. 4.

In what follows both the `freqtab()` and `kefreq()` functions will be used for illustrations using score data collected under different equating designs.

2.2.2 Score Data in the EG Design

The first illustration is for the case when we have a long vector of scores, one for each of the test forms. The length of each score vector equals the sample size of the group that is being administered the corresponding form. To exemplify how to create score frequency distributions using the `freqtab()` function, we use the

Math20EG data set. The scores in the Math20EG data set can be loaded and visualized using the following code

```
> data("Math20EG", package = "SNSequate")
> Math20EG
      x.freq y.freq
 [1,]      1      0
 [2,]      3      4
 [3,]      8     11
 [4,]     25     16
 [5,]     30     18
 [6,]     64     34
 [7,]     67     63
 [8,]     95     89
 [9,]    116     87
[10,]    124    129
[11,]    156    124
[12,]    147    154
[13,]    120    125
[14,]    129    131
[15,]    110    109
[16,]     86     98
[17,]     66     89
[18,]     51     66
[19,]     29     54
[20,]     15     37
[21,]     11     17
```

As seen from the output, the score data are already tabulated as frequencies. Each column contains the observed frequencies for each score in test form X and Y, respectively. The numbers in the first raw correspond to the observed frequencies for scores $X = 0$ and $Y = 0$, the ones in the second for $X = 1$ and $Y = 1$, and so on until the last raw that represents the frequencies for scores $X = 20$ and $Y = 20$. We expand this frequency data using the rep() function to obtain the whole vector of scores using the following code

```
> x.scores<-rep(0:20,Math20EG[,1])
> y.scores<-rep(0:20,Math20EG[,2])
```

The x.scores and y.scores are now score vectors of dimension equal to the sample sizes in forms X and Y, respectively. To create frequency distributions we write

```
> library(equate)
> eg.x<-freqtab(x.scores, 0:20)
> eg.y<-freqtab(y.scores, 0:20)
```

The objects eg.x and eg.y are now two-column matrices containing score frequency distributions. If instead of a long vector the score data are tabulated in frequencies, as is the case for the Math20EG data, the as.freqtab() function can be used to give the appropriate attributes to the object to be used in **equate**. For instance, using the frequencies in Math20EG and adding a vector of all possible total scores obtained (i.e., $\mathscr{X} = \{0, 1, \ldots, 19, 20\}$ in this case), the previously created `eg.x` and `eg.y` can equivalently be obtained by writing:

```
> eg.x<-as.freqtab(cbind(0:20,Math20EG[,1]))
> eg.y<-as.freqtab(cbind(0:20,Math20EG[,2]))
```

For test form X the output is

```
> eg.x
   total count
1      0     1
2      1     3
3      2     8
4      3    25
5      4    30
6      5    64
7      6    67
8      7    95
9      8   116
10     9   124
11    10   156
12    11   147
13    12   120
14    13   129
15    14   110
16    15    86
17    16    66
18    17    51
19    18    29
20    19    15
21    20    11
```

The `count` column contains the number of test takers obtaining each of the scores in the `total` column, and this column coincides with the `x.freq` column in the `Math20EG` data matrix.

A graphical representation of the score frequency distributions can be obtained in two ways. First, if the **equate** package has not been loaded, the generic `plot()` function in **R** can be used to visualize a scatter plot of scores vs frequencies. If an object of class `freqtab` is passed to `plot()`, then with the **equate** package loaded, a plot of frequencies as vertical lines will be displayed. This situation is shown in Fig. 2.1 which was produced using the following code

```
> par(mfrow=c(1,2))
> plot(0:20, Math20EG[, 1], pch = 0, ylab = "Frequency",
+ xlab = "Score")
> plot(eg.x)
```

Next, we illustrate how to create frequency distributions starting from 0-1 matrices of item responses (response patterns), using the `ADM1` admissions test data for the first administration (Y) and `ADM2` for the second (X). Assuming that the working directory has been properly set (see Sect. A.3), loading the data sets in the **R** session is done by writing

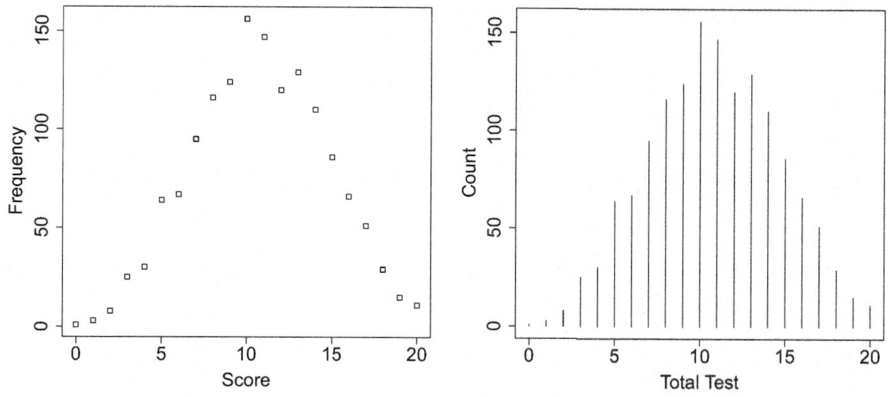

Fig. 2.1 Frequency distribution of X scores for the `Math20EG` data set

```
> load("ADM1.Rda")
> load("ADM2.Rda")
```

If an internet connection is available, an alternative and useful way to automatically load the data from the book's webpage is

```
> load(url("http://www.mat.uc.cl/~jorge.gonzalez/
+ EquatingRbook/ADM1.Rda"))
> load(url("http://www.mat.uc.cl/~jorge.gonzalez/
+ EquatingRbook/ADM2.Rda"))
```

Consider the data set `ADM1` which contains the answers to both the verbal and quantitative sections of the first test administration. Each row of this $10,000 \times 160$ binary matrix is the response pattern of a test taker. The first 80 columns correspond to answers to the quantitative test, and the remaining 80 are the answers to the verbal test. The data set `ADM2` which contains the answers for the second administration has similar characteristics. Suppose we want to equate the quantitative test such that scores on the second administration are equated to the scale of the first. We label the test form in the second administration X and the test form in the first administration Y. Our aim is to find the equating transformation $\varphi(x)$. Because the first 80 columns are the answers for the quantitative test, we use this part of the data set to obtain quantitative sum scores.

```
> quant.x<-apply(ADM2[,1:80],1,sum)
> quant.y<-apply(ADM1[,1:80],1,sum)
```

The objects `quant.x` and `quant.y` are now vectors of sum scores with length equal to the sample sizes in X and Y ($10,000$ in this case). For each row, the `apply()` function sums over the columns of the 0-1 matrix to obtain sum scores. Thus, score frequency distributions can be created using the following code.

```
> egADM.x<-freqtab(quant.x,0:80)
> egADM.y<-freqtab(quant.y,0:80)
```

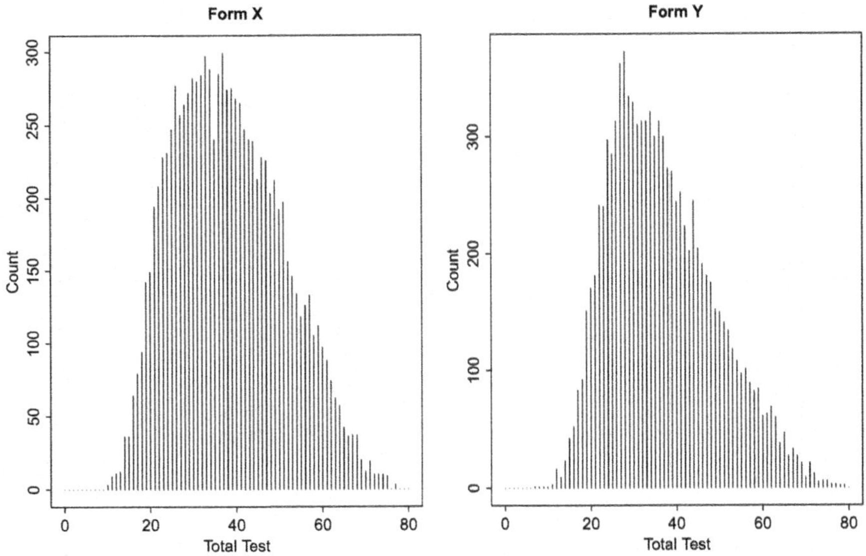

Fig. 2.2 Frequency distribution of X and Y scores for the ADM data under an EG design

A graphical representation of the score frequency distributions is shown in Fig. 2.2, which was created using the following code

```
> par(mfrow=c(1,2))
> plot(egADM.x,main="Form X")
> plot(egADM.y,main="Form Y")
```

When using **kequate**, the score data must first be converted into frequencies for each combination of score values. The `kefreq()` function is designed to create both univariate and bivariate score frequency distributions. To use `kefreq()` under the EG design, only two arguments need to be defined: `in1` and `xscores`. For instance, using the observed scores stored in the `quant.x` and `quant.y` objects created above

```
> library(kequate)
> egADMk.x <- kefreq(in1=quant.x,xscores=0:80)
> egADMk.y <- kefreq(quant.y,0:80)
```

The observed test scores are input in `in1` and the possible test score values are input in `xscores`. Note that the way `egADMk.y` was obtained exemplifies that shorthand writings are possible. In this case, it is not necessary to write neither of the argument names (`in1` and `xscores`).

The resulting objects `egADMk.x` and `egADMk.y` are two-column matrices containing the possible test score values (`X`, first column) and a vector with the frequencies for each test score value for test X (`frequency`, second column). The first 10 rows for `egADMk.y` are shown below

```
> head(egADMk.y, n=10)
   X frequency
1  0         0
2  1         0
3  2         0
4  3         0
5  4         0
6  5         0
7  6         0
8  7         1
9  8         1
10 9         1
```

Additional examples of preparing score distributions under the EG design using the ADM data can be found in the book's webpage.

2.2.3 *Score Data in the SG Design*

The `Math20SG` data are used for illustrating score data in the SG design. As indicated in Chap. 1, the SG design will produce a bivariate vector of scores. The data matrix thus contains the (joint) bivariate sample frequencies for X in rows and for Y in columns. The `Math20SG` data set has the following appearance

```
> data("Math20SG", package = "SNSequate")
> colnames(Math20SG)<-as.character(0:20)
> rownames(Math20SG)<-as.character(0:20)
> Math20SG

   0 1 2 3  4  5  6  7  8  9 10 11 12 13 14 15 16 17 18 19 20
0  0 0 1 0  0  0  0  0  0  0  0  0  0  0  0  0  0  0  0  0  0
1  0 0 1 0  1  1  0  0  0  0  0  0  0  0  0  0  0  0  0  0  0
2  0 0 1 0  2  1  3  0  0  1  0  0  0  0  0  0  0  0  0  0  0
3  0 0 1 5  6  3  8  1  1  0  0  0  0  0  0  0  0  0  0  0  0
4  0 0 2 7  4  6  4  3  1  3  0  0  0  0  0  0  0  0  0  0  0
5  0 0 3 3  5 12 14  8  9  6  3  1  0  0  0  0  0  0  0  0  0
6  0 0 1 4 10  9 12  9  8 10  4  0  0  0  0  0  0  0  0  0  0
7  0 0 1 3  5  7 16 16 11 17 10  5  3  0  1  0  0  0  0  0  0
8  0 0 1 1  3  8 16 14 12 24 20 11  3  3  0  0  0  0  0  0  0
9  0 0 0 1  3  4  8 19 20 17 17 13 11  9  2  0  0  0  0  0  0
10 0 0 0 0  1  2  6 14 20 19 28 24 17 11  9  3  2  0  0  0  0
11 0 0 0 0  1  3  3  6 13 17 21 23 27 14 13  2  2  1  1  0  0
12 0 0 0 0  0  1  0  5 11 14 16 26 18 11 10  3  3  1  1  0  0
13 0 0 0 0  0  0  1  4  8  8 20 21 19 16 13  9  6  3  1  0  0
14 0 0 0 0  0  0  0  1  4  3  3 17 18 26 11 21  4  1  1  0  0
15 0 0 0 0  0  0  1  0  1  3  4 10 12 15 15 10 10  3  1  1  0
16 0 0 0 0  0  0  0  0  0  1  1  1 11 12  8 13 10  7  1  1  0
17 0 0 0 0  0  0  0  0  0  0  2  1  5  4  8  9 11  5  3  3  0
18 0 0 0 0  0  0  0  0  0  0  0  0  0  5  0  4  4 11  4  1  0
19 0 0 0 0  0  0  0  0  0  0  0  0  1  1  2  2  2  3  3  1  0
20 0 0 0 0  0  0  0  0  0  0  0  0  1  0  0  0  2  3  3  2  0
```

The output shows that, for instance, 10 test takers obtained $X = 6$ *and* $Y = 4$ whereas only one test taker obtained $X = 1$ *and* $Y = 4$. To obtain the marginal score distributions, we need to sum over either row or columns depending on the marginal distribution we are interested in. The following code can be used to obtain the marginal frequency distributions of X and Y

```
> Math20SG.x<-rep(0:20,apply(Math20SG,1,sum))
> Math20SG.y<-rep(0:20,apply(Math20SG,2,sum))
```

Preparing bivariate score frequency distributions using **equate** is simple if one uses the `freqtab()` function as follows

```
> sg.data<-freqtab(cbind(Math20SG.x, Math20SG.y),
+ scales = list(0:20, 0:20), design="sg")
```

The arguments are now a two-column matrix of scores in test X (first column) and Y (second column), along with a list of vectors containing the score scales for each column in the matrix.

For the admissions test data, the data set `ADM12` contains the answers for 8,000 test takers who have results on both test forms X and Y for the quantitative and verbal sections. The data matrix has 320 columns where the first 80 are answers to the quantitative test in X, the second 80 are answers for the verbal test in X, the third 80 are answers for the quantitative test in Y, and the last 80 are the answers for the verbal test in Y. If we want to equate the quantitative test under the SG design, we first need to obtain sum scores in the following way

```
> load("ADM12.Rda")
> sgADM.x <- apply(ADM12[,1:80],1,sum)
> sgADM.y <- apply(ADM12[,161:240],1,sum)
> sgADM.data <- freqtab(cbind(sgADM.x, sgADM.y),
+ scales = list(0:80, 0:80))
```

The resulting `sgADM.data` object is a 81 × 81 matrix of bivariate frequencies. The way in which the `sgADM.data` object is displayed in the output, however, corresponds to a three-column array where the first and second column contain all possible combinations of score pairs and the third column is the observed frequency of such score pair combinations. For instance the following portion of output

```
     total anchor count
....    ..     ..     .
....    ..     ..     .
6059    64     74     1
6060    65     74     0
6061    66     74     2
6062    67     74     1
6063    68     74     2
....    ..     ..     .
....    ..     ..     .
```

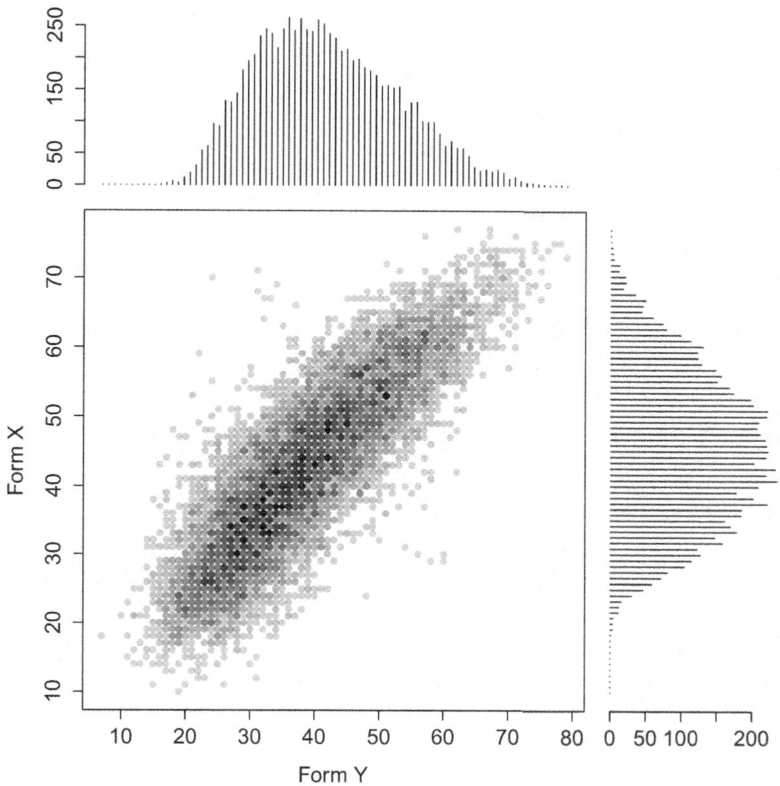

Fig. 2.3 Marginals of the bivariate score frequency distributions for the ADM data under the SG design

shows that only one test taker scored 64 on test form X *and* 74 on test form Y. Also, there were no test takers scoring 65 on test form X and 74 on test form Y, and only two test takers scored 68 on test form X and 74 on test form Y, and so forth. When working with bivariate frequency distributions, the `plot()` function will draw marginal score distributions for X and Y as well as the score points for both test forms as shown in Fig. 2.3. Note that when applied to a bivariate frequency distribution created using `freqtab()`, the default option for `plot()` is for the case when a NEAT design is assumed (see the next section). This means that the function will assign labels for anchor (Y axes) and total scores (X axes), respectively. Thus, to obtain a plot with correct labels in the case of an SG design the code can slightly be modified in its arguments by writing `plot(sgADM.data,xlab='Form Y',ylab='Form X')`.

A similar code that produces the same results as the one just shown for `freqtab()` can be written using the `kefreq()` function in **kequate**. A notable difference is that the obtained output is now a data frame, which permits one to filter the results as desired. For instance, in order to obtain the portion of output shown above for the `freqtab()` function, we can write

```
> sgADMxy <- kefreq(sgADM.x,0:80,sgADM.y,0:80)
> sgADMxy[6059:6063,]
      X  A frequency
6059 64 74         1
6060 65 74         0
6061 66 74         2
6062 67 74         1
6063 68 74         2
```

The `kefreq()` function internally handles all of the different combinations of the score frequencies (`frequency`) with each combination of score values, for test form X (`X`) and for test form Y (denoted `A` in the output).

2.2.4 Score Data in the NEAT Design

Similarly to the SG design, collecting data under the NEAT design will produce bivariate score data. The difference is that under the NEAT design, the bivariate scores vectors will be the pairs of total test scores (in the first coordinate) and total anchor scores (in the second coordinate). The examples given in the help section for the `freqtab()` function in **equate** use the `KB36` data and show how to create frequency distributions using the scores frequencies. Although the same data are used here, we will illustrate how to start from the 0-1 matrix, and show how to obtain the same bivariate frequency score distributions. First, the `KBneatX` and `KBneatY` data are loaded[2]

```
> load("KBneatX.Rda")
> load("KBneatY.Rda")
```

The `freqtab()` function is used with the argument `items` to indicate which items should be summed to obtain scores for both the total test and the anchor test. The argument `items` is a list whose first element indicates the range of items to be summed in order to obtain the total test score. The second element in the list indicates the items that should be summed to obtain the total anchor scores. Note that when anchor scores are considered internal, the total score is considered to be the sum of "unique" correctly answered items plus the sum of the corresponding anchor score items. For instance, for the `KB36` data, both forms X and Y are built using 24 unique items for each form and 12 anchor items that are common to both forms. The total score is thus obtained by summing the 24 plus 12 items for each form. As mentioned previously in this chapter, every third items in the `KB36` data is a common item, thus one way to obtain the test items to be summed is as follows

[2]Note that the 0-1 response matrices for forms X and Y can also be obtained from the `KB36` list by writing `KB36$KBformX` and `KB36$KBformY`, respectively.

```
> KB.neat.x <-freqtab(KBneatX[,1:36], items =
+ list(1:36,seq(3, 36, 3)), scales=list(0:36, 0:12))
> KB.neat.y <-freqtab(KBneatY[,1:36], items =
+ list(1:36,seq(3, 36, 3)), scales=list(0:36, 0:12))
```

Using the following code, the resulting `KB.neat.x` and `KB.neat.y` objects can be shown to give the same outputs obtained using the example code provided in the help section for `freqtab()` in **equate**

```
> KB.neat.x2 <- freqtab(KBneat$x, scales =
+ list(0:36, 0 :12))
> KB.neat.y2 <- freqtab(KBneat$y, scales =
+ list(0:36, 0 :12))
> all.equal(KB.neat.x,KB.neat.x2)
[1] TRUE
> all.equal(KB.neat.y,KB.neat.y2)
[1] TRUE
```

The resulting objects are 37×3 matrices of bivariate frequencies. The displayed output for the first 10 scores in form X is

```
> head(KB.neat.x,n=10)
   total anchor count
1      0      0     0
2      1      0     0
3      2      0     1
4      3      0     2
5      4      0     0
6      5      0     2
7      6      0     0
8      7      0     6
9      8      0     0
10     9      0     3
```

We will now illustrate how to prepare score distributions under the NEAT design using the `ADM` data set. This time, the verbal section of the test will be used for illustration. For test forms X and Y the `ADMneatX` and `ADMneatY` data sets contain 2,000 test takers who were given a verbal anchor test composed of 40 items. Both data sets are a 2000×120 matrix where the first 40 columns contain the answers to the verbal anchor items and the remaining 80 columns are the answers to the verbal items in the corresponding test form. After loading the data, we first create vectors of both anchor and items sum scores

```
> load("ADMneatX.Rda")
> load("ADMneatY.Rda")
> verb.xa <- apply(ADMneatX[,1:40],1,sum)
> verb.ya <- apply(ADMneatY[,1:40],1,sum)
> verb.x  <- apply(ADMneatX[,41:120],1,sum)
> verb.y  <- apply(ADMneatY[,41:120],1,sum)
```

Before creating the frequency distributions, we need to build two-column matrices with the test sum scores in the first column and the anchor sum scores in the second column. Once these matrices are obtained, the `freqtab()` function

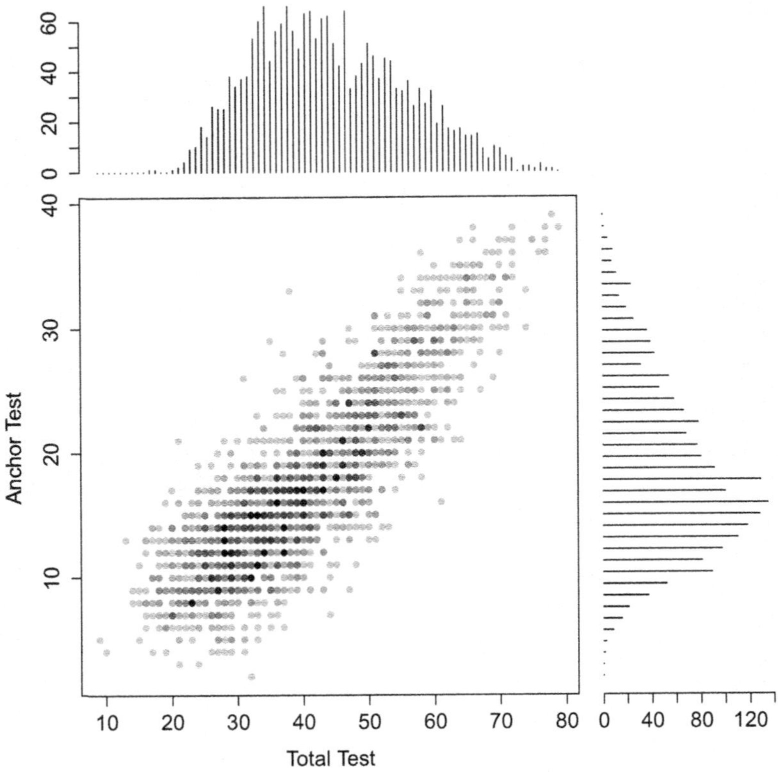

Fig. 2.4 Marginals of the bivariate distribution of (X, A) for the ADM data

is used to produce the bivariate score frequency distributions that are needed under the NEAT design. The following code is used to perform these tasks.

```
> neat.X <-cbind(verb.x,verb.xa)
> neat.Y <-cbind(verb.y,verb.ya)

> neat.x1<-freqtab(x = neat.X, scales =
+ list(0:80, 0:40))
> neat.y1<-freqtab(x = neat.Y, scales =
+ list(0:80, 0:40))
```

The `plot()` function in this case will produce a graphical representation of total test scores and anchor scores as seen in Fig. 2.4.

When using the **kequate** package, the following code serves to prepare the admissions test data

```
> neatk.x <- kefreq(in1 = verb.x, xscores = 0:80,
+ in2 = verb.xa, ascores = 0:40)
> neatk.y <- kefreq(verb.y, 0:80, verb.ya, 0:40)
```

Score frequency distributions, as illustrated for the CB and NEC designs can also be obtained with the **kequate** function `kefreq()`. How this should be done is illustrated in Chap. 4. Once the test score distributions are obtained and prepared for use in **R**, the next step is either to perform the equating of interest directly or to first presmooth the score distributions to repair irregularities in the observed score distributions. Some statistical models for presmoothing are described in the following section.

2.3 Presmoothing the Score Distributions

When collecting equating data, samples from the populations of interest are drawn according to a certain design. The resulting data are either pairs of independent score random variables (e.g., for the EG design) or bivariate random vectors (e.g., for the SG, CB, and NEAT designs) (see Sect. 1.3.1). Intuitively, the larger the sample size, the better the score distribution will represent the population. However, due to sampling error, irregularities are often seen in these distributions, and good *estimates* of the discrete distribution are needed. A common way to reduce these irregularities is to *presmooth* the score distributions obtained from a particular data collection design. Mathematically, presmoothing means modeling the score distributions by relating the score probabilities with the model parameters. Formally, if **p** represents the score probabilities under some data collection design in equating, and $\mathscr{M}(\boldsymbol{\theta})$ is a model parameterized by $\boldsymbol{\theta}$, then the model

$$\mathbf{p} = \mathscr{M}(\boldsymbol{\theta}) \tag{2.1}$$

is fitted to the data and the model parameter estimates $\hat{\boldsymbol{\theta}}$ are used to determine $\hat{\mathbf{p}}$. Note that under the EG design we have $\mathbf{p} = \Pr(X = x_j) = p_j$, whereas under the SG and NEAT designs $\mathbf{p}$ equals $\Pr(X = x_j, Y = y_k) = p_{jk}$ and $\Pr(X = x_j, A = a_l) = p_{jl}$, respectively. In the case of the CB design, the two independent bivariate probabilities are specified as $\Pr(X_1 = x_{1j}, Y_2 = y_{2k}) = p_{(12)jk}$ and $\Pr(X_2 = x_{2j}, Y_1 = y_{1k}) = p_{(21)jk}$, respectively. Similar score probabilities can be defined for Y under the EG and NEAT designs. Although various options for $\mathscr{M}(\boldsymbol{\theta})$ are possible, in this section we discuss the log-linear models that are included in the `presmoothing()` function in the **R** package **equate**. In Chap. 7, the use of item response theory models is discussed for the estimation of **p**.

2.3.1 Polynomial Log-Linear Models for Presmoothing

Because each score is obtained with a certain probability, the main idea in log-linear presmoothing is to assume that observed score frequencies, each with a specific probability of occurrence, are realizations of a multinomial distribution. More specifically, let j represent the test scores and n_j the frequency of test score j. If $p_j = \Pr(X = x_j)$ and $\mathbf{n} = (n_1, \ldots, n_J)$, then $\mathbf{n} \sim \text{Multinomial}(\mathbf{n}, \mathbf{p})$. The log-likelihood function in this case is

$$\ell(p_j) = \sum_j n_j \log(p_j)\,. \tag{2.2}$$

Thus, assuming a log-linear model for p_j of the form $\log(p_j) = \beta_0 + \mathbf{b}_j'\boldsymbol{\beta}$, where β_0 is a normalization constant, $\mathbf{b}_j'$ is a vector of known constants, and $\boldsymbol{\beta}$ is a vector of parameters, the likelihood equation to obtain $\hat{\boldsymbol{\beta}}$ implies that

$$\sum_j b_{ij}(n_j/N) = \sum_j b_{ij}\hat{p}_j\,, \tag{2.3}$$

where N is the sum of the observed score frequencies. Equation 2.3 shows what is known as the *moment-matching property* of log-linear models indicating that sample and fitted moments are matched. Although other moments are possible to be used (see e.g., von Davier et al. 2004, Appendix C), power moments are mostly used and will be used here for illustrations. Under this specification, $b_{ij} = x_j^i$, and the polynomial log-linear models used to model p_j becomes

$$\log(p_j) = \beta_0 + \sum_{i=1}^{T_r} \beta_i (x_j)^i = \beta_0 + \beta_1 x_j + \beta_2 x_j^2 + \ldots + \beta_{T_r} x_j^{T_r}. \tag{2.4}$$

where T_r denotes the highest polynomial degree, $i = 1, \ldots, T_r$. A similar model can be specified using the Y scores. For designs where bivariate distributions are involved, a bivariate log-linear model can be fitted to the score data. Besides the marginal moments of X and Y, interaction terms can be included in the model. For instance, let j and k represent the test scores. Under the SG design we have $p_{jk} = \Pr(X = x_j, Y = y_k)$ and thus the model becomes

$$\log(p_{jk}) = \beta_0 + \sum_{i=1}^{T_r} \beta_i^X (x_j)^i + \sum_{i=1}^{T_s} \beta_i^Y (y_k)^i + \sum_{i=1}^{L_r}\sum_{l=1}^{L_s} \beta_{il}^{XY} x_j^i y_k^l, \tag{2.5}$$

where T_s denotes the highest polynomial degree in y_k ($i = 1, \ldots, T_s$) and T_r is defined as before. L_r and L_s are the sum limits that serve to accommodate cross moments in the models. The superscripts X, Y, and XY are used to differentiate between the β parameters accompanying the powers of the x_j scores, the y_k scores, and the cross products of the scores $x_j y_k$, respectively. For a more detailed review on the use of loglinear models to describe discrete test score distributions, refer to Rosenbaum and Thayer (1987), Holland and Thayer (1987, 2000), Moses and Holland (2009).

2.3.2 *Polynomial Log-Linear Smoothing in equate*

To fit polynomial loglinear models in the **equate** package, the `presmoothing()` function is used. Its main arguments and a typical call are as follows

```
presmoothing(x, smoothmethod, scorefun, degrees, ...)
```

The `x` argument is most typically a previously created object of class "freqtab", which specifies a univariate or multivariate score distribution. Writing `smoothmethod = "loglinear"`, makes a call to the general **R** function `glm()` to fit generalized linear models using a Poisson family so that log-linear models are obtained. Depending on whether a univariate or bivariate model is fitted, different possibilities to specify the covariates in the model (i.e., the x_j^i), are available. The first possibility is to use a matrix of score functions (`scorefun`) where each column is a predictor in the model of interest. The second possibility is to include the highest desired polynomial terms in the argument `degrees` for univariate or bivariate moments. For instance, to fit the model in Eq. (2.4) with $T_r = 2$, `scorefun` will contain a matrix having x_j and x_j^2 as columns. If, the argument `degrees` is used instead, then `degrees=2` will produce the same covariates in the model. If a bivariate model such as in Eq. (2.5) is fitted, for instance for $T_r = L_r = 2$ and $T_s = L_s = 3$ then `scorefun` will contain a matrix having $x_j, x_j^2, y_k, y_k^2, y_k^3, x_jy_k, x_jy_k^2, x_jy_k^3, x_j^2y_k, x_j^2y_k^2$, and $x_j^2y_k^3$ as columns, whereas if the `degrees` argument is used, then one might specify `degrees=list(2,3)`. The first integer in the list indicates that the first variable, x, is modeled up to the second power whereas the second, indicates that y is modeled up to the third power. Log-linear models fitted using this option are considered hierarchical so that all lower-order terms are included in the model (Fienberg 1980). It is, however, also possible to fit a model with no hierarchical structure. To exemplify this, consider a model with $T_r = 2$, $T_s = 3$, $L_r = 1$ and $L_s = 2$ so that the model equation becomes

$$\begin{aligned}\log(p_{jk}) &= \beta_0 + \beta_1^X(x_j) + \beta_2^X(x_j)^2 \\ &\quad + \beta_1^Y(y_k) + \beta_2^Y(y_k)^2 + \beta_3^Y(y_k)^3 \\ &\quad + \beta_{11}^{XY}(x_j)(y_k) + \beta_{12}^{XY}(x_j)(y_k)^2. \end{aligned} \tag{2.6}$$

In this case we write `degrees=list(c(2,3),c(1,2))`. The integer values in the vector `c(2,3)` mean that x and y are modeled up to the second and third power, respectively, and `c(1,2)` means that in the interaction only the main effect of x is modeled whereas for y the effects up to the second power are used. Note that, because `freqtab()` is capable of producing multivariate score frequency distributions using more than just two variables, a general use of `degrees` implies that the given list contains as many vectors as there are variables being modeled. The order in which such vectors appear in the list corresponds to main effects, bivariate interaction effects, trivariate interaction effects, and so on. Thus, if x, y, and z are three score variables, a model including up to the third power for each univariate distribution, up to the second power for each two-way interaction, and up to the second power for the three-way interaction is fitted by setting `degrees=list(c(3,3,3),c(2,2,2),c(2,2,2))`.

A third option for fitting polynomial log-linear models is to use the argument `grid`. This argument receives as input a matrix with as many rows as covariates (main effects and interactions) that there are in the model. In each row-vector, whose length is equal to the number of variables being modeled, the corresponding power degree for the variables must be specified. For instance, to fit the model in Eq. (2.6), `grid` will be a 7×2 matrix of the form

$$\begin{pmatrix} 1 & 0 \\ 2 & 0 \\ 0 & 1 \\ 0 & 2 \\ 0 & 3 \\ 1 & 1 \\ 1 & 2 \end{pmatrix}$$

which can easily be written as

```
> grid = matrix(c(1, 0, 2, 0, 0, 1, 0, 2, 0, 3, 1, 1,
+ 1, 2), ncol = 2, byrow = T)
```

In what follows, the use of the `presmoothing()` function is exemplified for both univariate and bivariate score distributions.

2.3.3 *Examples*

2.3.3.1 Smoothing Univariate Distributions

The EG design will provide univariate score vectors and this type of data is used here to illustrate univariate presmoothing with log-linear models. For this example, score frequencies are taken from Table 7.1 in von Davier et al. (2004), which are stored in the `Math20EG` object described previously. We will reproduce the results of Table 7.2 in von Davier et al. (2004) where the fitted frequencies are obtained by fitting polynomial log-linear models that preserve the second and third moments of X and Y, respectively

$$\log(p_j) = \beta_0^X + \beta_1^X x_j + \beta_2^X x_j^2, \tag{2.7}$$

$$\log(p_k) = \beta_0^Y + \beta_1^Y y_k + \beta_2^Y y_k^2 + \beta_3^Y y_k^3. \tag{2.8}$$

We can use the eg.x and eg.y frequencies created in Sect. 2.2.2 and the following code

```
> tab72.x <- presmoothing(eg.x,smoothmethod =
+ "loglinear", degrees = 2,asfreqtab=FALSE)
> tab72.y <- presmoothing(eg.y,smoothmethod =
+ "loglinear", degrees = 3,asfreqtab=FALSE)
```

which produces the following output that coincides with the fitted values as shown in Table 7.2 in von Davier et al. (2004).

```
> round(cbind(tab72.x,tab72.y),2)
   tab72.x tab72.y
0     3.30    1.71
1     6.44    3.77
2    11.77    7.65
3    20.17   14.24
4    32.43   24.44
5    48.89   38.75
6    69.10   56.98
7    91.57   77.91
8   113.79   99.35
9   132.58  118.54
10  144.83  132.72
11  148.36  139.87
12  142.49  139.15
13  128.32  131.10
14  108.35  117.31
15   85.79  100.00
16   63.69   81.46
17   44.33   63.60
18   28.93   47.73
19   17.71   34.54
20   10.16   24.18
```

A plot showing the original score frequency distributions for tests forms X and Y and the result of presmoothing them using the log-linear models in the example above is shown in Fig. 2.5. The figure was produced with the code

```
> par(mfrow=c(1,2))
> plot(eg.x, y = tab72.x, legendtext = "degree = 2")
> plot(eg.y, y = tab72.y, legendtext = "degree = 3")
```

2.3.3.2 Smoothing a Bivariate Distribution

Both SG and NEAT designs produce bivariate data. To illustrate how to perform a log-linear smoothing with the **equate** package, the Math20SG test data are used with the sg.data object created in Sect. 2.2.3. Suppose we want to fit the loglinear model in Eq. (2.5) with $T_r = T_s = 3$ and to consider only the main effects of score

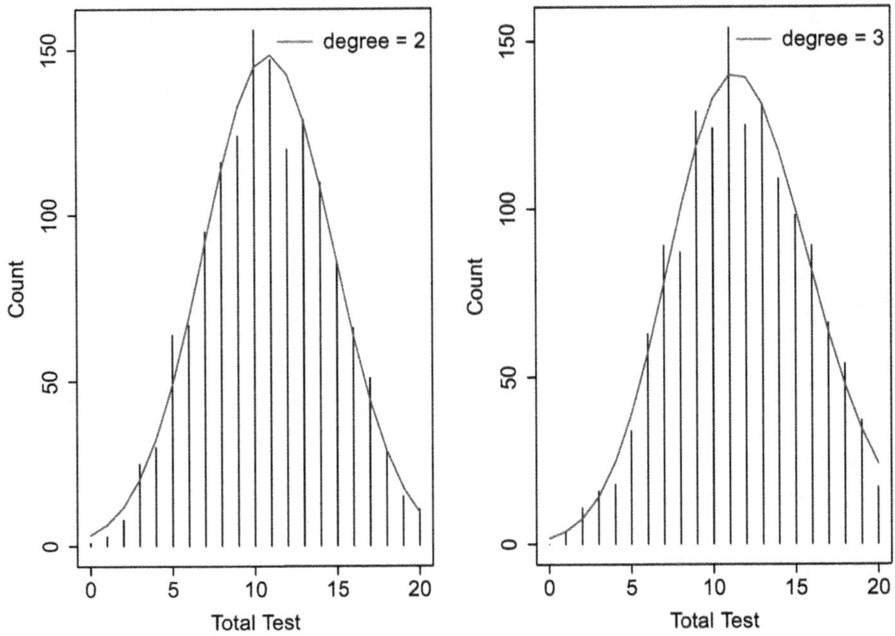

Fig. 2.5 Original (bars) and smoothed (continuous line) score frequency distributions for test forms X (*left panel*) and Y (*right panel*) for the `Math20EG` data

covariates so that $L_r = Ls = 0$ and thus $\sum_{i=1}^{L_r}\sum_{l=1}^{L_s}\beta_{il}^{XY}x_j^i y_k^l = 0$. The following code can be used to fit this model

```
> sg.smooth<-presmoothing(sg.data, smoothmethod =
+ "loglinear", degrees = list(c(3, 3),c(0, 0)),
+ showWarnings = FALSE)
> plot(sg.data,sg.smooth,ylab="Form Y",xlab="Form X",
+ addlegend=FALSE)
```

A plot showing the original discrete distributions and the effect of presmoothing is shown in Fig. 2.6. An example of bivariate presmoothing under the NEAT design using the `KB36` data can be found in Albano (2016). More examples of presmoothing are given on the book's webpage.

2.3.4 *Choosing the Best Log-Linear Model*

Choosing the appropriate polynomial degree in presmoothing reduces to a comparison of different log-linear models. This task can be performed using the `compare=TRUE` option in the `presmoothing()` function. The function will fit all nested models with the maximum degree as specified. The output is an ANOVA table containing information of fit statistics such as the Akaike Information

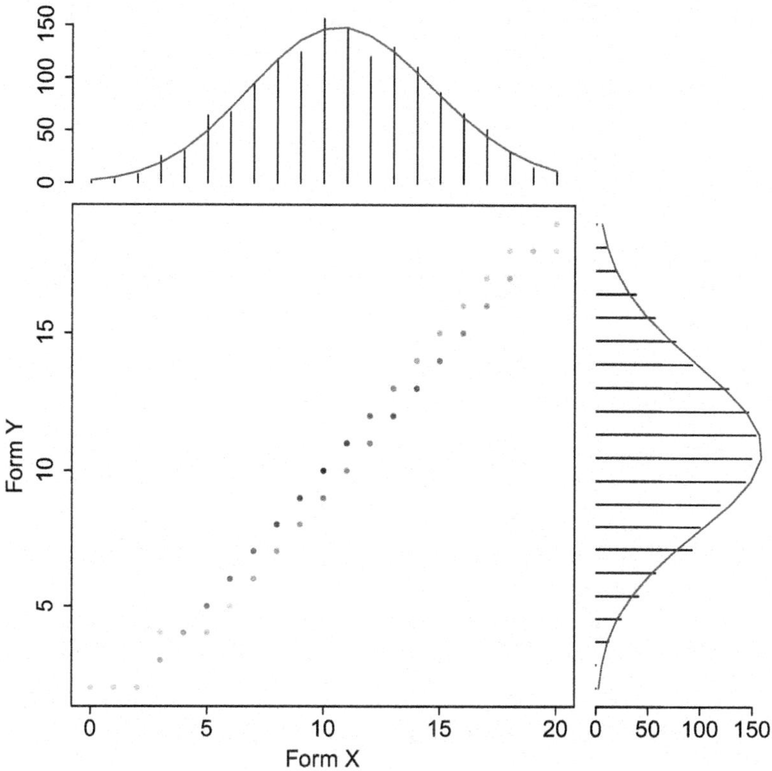

Fig. 2.6 Marginals of the bivariate distribution: SG design `Math20SG` data set

Criterion (AIC) (Akaike 1981), the Bayesian Information Criterion (BIC) (Schwarz 1978), and the likelihood ratio χ^2 test. The smallest values of AIC and BIC lead to a preferable model.

To illustrate how to obtain AIC and χ^2 measures, we reproduce parts of Table 3.1 in Kolen and Brennan (2014) using the `ACTmath` data. The table shows values for AIC, χ^2, and the difference between two χ^2 values. First, score frequency distributions are created as follows

```
> xACT<-as.freqtab(ACTmath[,c(1,2)])
> yACT<-as.freqtab(ACTmath[,c(1,3)])
```

The `compare=TRUE` argument is used to indicate that all (nested) models with polynomial degree less than 10 are to be fitted.

```
> KBtab31<-presmoothing(xACT,smoothmethod="loglinear",
+ degrees = 10, compare=TRUE)
```

Among other information, the `KBtab31` object contains fit indexes such as the AIC and χ^2 statistics. Because the output does not give the χ^2 differences directly, we create an auxiliary matrix `aux`, that will produce the desired differences when multiplied by the χ^2 values ordered in a vector. The values for χ^2, their differences, and AIC are stored in the object `fit.measures` which is used to produce Table 2.1. The following code is used to carry out these tasks

```
> aux<-matrix(c(
+ 1,-1, 0, 0, 0, 0, 0, 0, 0, 0,
+ 0, 1,-1, 0, 0, 0, 0, 0, 0, 0,
+ 0, 0, 1,-1, 0, 0, 0, 0, 0, 0,
+ 0, 0, 0, 1,-1, 0, 0, 0, 0, 0,
+ 0, 0, 0, 0, 1,-1, 0, 0, 0, 0,
+ 0, 0, 0, 0, 0, 1,-1, 0, 0, 0,
+ 0, 0, 0, 0, 0, 0, 1,-1, 0, 0,
+ 0, 0, 0, 0, 0, 0, 0, 1,-1, 0,
+ 0, 0, 0, 0, 0, 0, 0, 0, 1,-1),
+ ncol=10,byrow=TRUE)

> fit.measures<-cbind(Chi2=KBtab31[[2]],DifChi2=
+ c(aux%*%KBtab31[[2]],NA), AIC =
+ KBtab31[[2]]+2*c(2,3,4,5,6,7,8,9,10,11))
```

From Table 2.1 one can see that a model with $T_r = 6$ produces the smallest AIC and thus is the preferred model.

A graphical way to compare the different models is to add the `stepup` argument in the `presmoothing()` function and then to plot the original vs the smoothed score distributions.

```
> KBtab31<-presmoothing(xACT,smoothmethod="loglinear",
+ degrees = 10, stepup = TRUE)
> plot(xACT,KBtab31)
```

Setting `stepup=TRUE` will result in fitting a number of nested models. For the current example, ten log-linear models will be fitted, starting with the one that considers one polynomial degree ($T_r = 1$), followed by a model fitting two polynomial degrees ($T_r = 2$) up until a model considering ten polynomial degrees ($T_r = 10$). Figure 2.7 shows the graphical comparison and can be contrasted with the results in Table 2.1.

Table 2.1 Fit indices for various polynomial log-linear models

T_r	χ^2	$\chi^2_C - \chi^2_{C+1}$	AIC
1	2215.02	1769.83	2219.02
2	445.19	232.36	451.19
3	212.82	172.03	220.82
4	40.80	5.01	50.80
5	35.78	5.18	47.78
6	30.61	0.20	44.61
7	30.40	0.46	46.40
8	29.94	0.03	47.94
9	29.91	0.23	49.91
10	29.68	NA	51.68

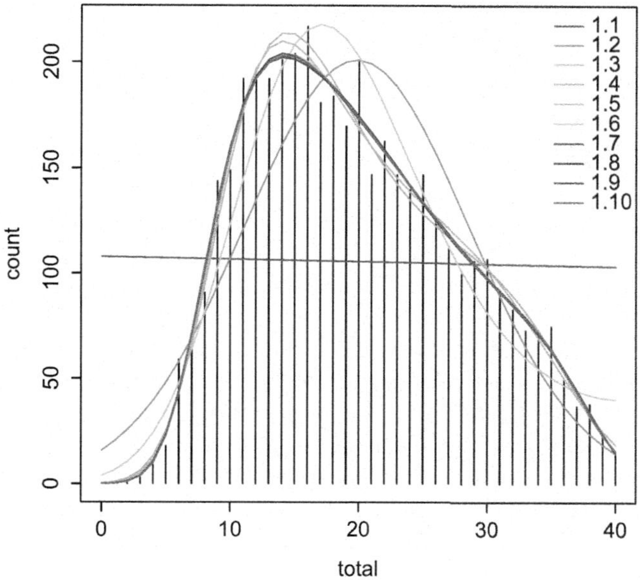

Fig. 2.7 Score frequency distribution (bars) and smoothed frequencies for ten different log-linear models

2.4 Using Other Arguments, Packages and Functions

In the previous sections we have shown how to perform presmoothing by fitting loglinear models to the data. The `presmoothing()` function also supports alternative methods for presmoothing that make adjustments to scores with low or zero frequencies. For instance setting `smoothmethod = "bump"` will add the proportion `jmin`, to be specified by the user, to each score point and then will adjust the probabilities to sum to 1. If `smoothmethod = "average"` then frequencies falling below the minimum `jmin` will be replaced with averages of adjacent values as described by Moses and Holland (2008). Useful examples can be found in the help section for the `presmoothing()` function in **equate**.

If only loglinear models are used for the presmoothing, then instead of using `presmoothing(x, smoothmethod= "loglinear")` one can obtain the same results using the function `loglinear()`. For instance, if one wants to fit the first two moments of X for the `Math20EG` data, the code

```
> loglinear(eg.x, degrees = list(2, 0, 0))
```

will produce exactly the same results as the ones shown in Sect. 2.3.3.

The general **R** function `glm()` can be used directly to presmooth the data, and output objects from this function are typically used when presmoothed data are needed in **kequate**. This will be discussed and exemplified in detail in Chap. 4.

Finally, in **SNSequate** the function `loglin.smooth()` also performs loglinear presmoothing. How this function should be used is illustrated in Chap. 4.

2.5 Summary

In this chapter we have shown how to prepare and read in score data under different data collecting designs in equating. We have also introduced presmoothing as an initial step in equating. Functions from the **equate** and **kequate** packages were used in the examples. In Chap. 1, the NEC design was also described. We have however, postponed the description of how data should be prepared under this design until Chap. 4, where we discuss this design in detail. Because presmoothing is a key step in kernel equating, as will be seen in Chap. 4, further examples using designs not considered here will be given in that chapter as well.

In the next chapter, various traditional methods of equating will be introduced. The score distributions derived in this chapter will serve as inputs for the examples using the `equate()` function from the **equate** package.

References

Akaike, H. (1981). Likelihood of a model and information criteria. *Journal of Econometrics, 16*(1), 3–14.

Albano, A. D. (2016). equate: An R package for observed-score linking and equating. *Journal of Statistical Software, 74*(8), 1–36.

Andersson, B., Bränberg, K., & Wiberg, M. (2013). Performing the kernel method of test equating with the package kequate. *Journal of Statistical Software, 55*(6), 1–25.

Fienberg, S. (1980). *The analysis of cross-classified, categorical data* (2nd ed.). Cambridge, MA: The MIT Press.

González, J. (2014). SNSequate: Standard and nonstandard statistical models and methods for test equating. *Journal of Statistical Software, 59*(7), 1–30.

Holland, P., & Thayer, D. (2000). Univariate and bivariate loglinear models for discrete test score distributions. *Journal of Educational and Behavioral Statistics, 25*(2), 133–183.

Holland, P. W., & Thayer, D. T. (1987). Notes on the use of log-linear models for fitting discrete probability distributions. *ETS Research Report Series, 1987*(2), i–40.

Kolen, M., & Brennan, R. (2014). *Test equating, scaling, and linking: Methods and practices* (3rd ed.). New York: Springer.

Moses, T., & Holland, P. (2008). Notes on a general framework for observed score equating. *ETS Research Report Series, 2008*(2), i–34.

Moses, T., & Holland, P. (2009). Selection strategies for univariate loglinear smoothing models and their effect on equating function accuracy. *Journal of Educational Measurement, 46*, 159–176.

Rosenbaum, P., & Thayer, D. (1987). Smoothing the joint and marginal distributions of scored two-way contingency tables in test equating. *British Journal of Mathematical and Statistical Psychology, 40*, 43–49.

Schwarz, G. (1978). Estimating the dimension of a model. *The Annals of Statistics, 6*, 461–464.

von Davier, A. A., Holland, P., & Thayer, D. (2004). *The kernel method of test equating*. New York: Springer.

Chapter 3
Traditional Equating Methods

Abstract This chapter describes traditional equating methods and their implementation in **R**. The **equate** package (Albano, J Stat Softw 74(8):1–36, 2016) will be used the most, although the possibility to use **SNSequate** (González, J Stat Softw 59(7):1–30, 2014) for traditional equating methods will also be explored. The methods included in this chapter are mean, linear and equipercentile equating for different data collection designs.

3.1 Equipercentile, Linear, and Mean Equating Transformations

Definition 1.1 given in Chap. 1 is general and it handles various types of equating transformations. The equipercentile equating transformation is formally defined exactly as presented in Definition 1.2. As described in Chap. 1, when using equipercentile equating, test form X is equated to test form Y by identifying X scores on test form X that have the same percentile ranks as Y scores on test form Y. The equipercentile equating transformation can be written in a closed form if we assume that X and Y are continuous random variables with cumulative distribution functions (CDF) F_X and F_Y, respectively,

$$\varphi(x) = F_Y^{-1}(F_X(x)). \tag{3.1}$$

All equating transformations can be shown to be special cases of Eq. (3.1). In particular, we have seen in Sect. 1.4 that linear equating is also a special case when the score distributions are location-scale families. Linear equating relies on the assumption that the differences in difficulty across test forms can be completely described by the first two moments of the score random variables (or of their distributions), i.e. the means (μ_X and μ_Y) and standard deviations (σ_X and σ_Y) coming from the test forms X and Y, respectively. The general definition of the linear equating function is

$$\varphi(x; \mu_X, \mu_Y, \sigma_Y, \sigma_X) = \frac{\sigma_Y}{\sigma_X}[x - \mu_X] + \mu_Y. \tag{3.2}$$

J. González, M. Wiberg, *Applying Test Equating Methods*,
Methodology of Educational Measurement and Assessment,
DOI 10.1007/978-3-319-51824-4_3

The mean equating method emerges for the case when only location families of score distributions for X and Y are concerned, or when the variance of both location-scale distributions are assumed to be equal (see Eq. (3.2)). Mean equating relies on the assumption that the test forms X and Y only differ in difficulty by their means (first moments) (μ_X and μ_Y) along the score scale

$$\varphi(x; \mu_X, \mu_Y) = X + \mu_Y - \mu_X. \tag{3.3}$$

It should be noted that mean equating is typically used only for illustrative purposes because it might be overly simplistic in many operating testing situations.

All equating transformations that are described in this chapter are based on Eqs. (3.1), (3.2), or (3.3). They will differ in the way the involved parameters in φ have to be estimated and in that different estimation procedures use different assumptions that depend on the data collection design which are described next.

3.2 Assumptions in the Different Designs

We have seen in Sect. 1.3.1 that different data collection designs for equating collect test score data using samples from either one common population, T, or two different populations P and Q. For instance, in the SG, EG, and CB designs, a unique sample, two independent samples, and a mixture of both samples from a common target population T are obtained and used for the estimation of φ (see Table 1.1). However, in both the NEAT and NEC designs, samples are taken from different populations, P and Q, so that a target population must be defined from these two populations. Because the definition of equating given in Definition 1.2 is established for a common population T, some assumptions will be needed to obtain a valid equating, especially when score data have been collected under the NEAT design. In what follows, the different assumptions within these designs are discussed.

3.2.1 Assumptions in EG, SG, and CB Designs

In the SG design, both test forms X and Y are administered to a unique sample group (G) of test takers. In the EG design, two independent groups (G_1 and G_2) of test takers are administered one of the two test forms X or Y. In the CB design, two independent groups are administered both test forms, but in different order. What is common in all of these designs is that samples are taken from a common (unique) population, T. This means that the univariate (EG) or bivariate (SG, CB) score distributions are directly obtained, and thus the parameters related to these distributions are directly estimable and can be used for the straightforward estimation of φ. For the EG, SG and CB designs we do not need any particular assumptions for making inferences on the equating transformation. Everything is directly observable and estimable because all data come from a common population.

3.2.2 Assumptions in the NEAT Design

When equating is performed under the NEAT design, the main difference compared with the EG, SG, and CB designs is that in the latter designs, samples are taken from a common population, T, whereas in the NEAT design, samples come from two different populations P and Q. Under the NEAT design, the target population is built as a weighted combination of P and Q, which is also referred to as a *synthetic* population (Braun and Holland 1982) and is defined as

$$T = w_P P + w_Q Q, \tag{3.4}$$

where it holds that $w_P + w_Q = 1$ and $w_P, w_Q \geq 0$.

Using the definition in Eq. (3.4), the corresponding score distributions of X and Y defined on the common target population, T, which we denote as $f_{XT}(x)$ and $f_{YT}(y)$, respectively, can be written accordingly as a weighted combination of the form

$$\begin{aligned} f_{XT}(x) &= w_P f_{XP}(x) + w_Q f_{XQ}(x), \text{ and} \\ f_{YT}(y) &= w_P f_{YP}(y) + w_Q f_{YQ}(y). \end{aligned}$$

Here, $f_{XP}(x)$ and $f_{XQ}(x)$ are the distributions of X scores in P and Q, respectively, and similarly, $f_{YP}(y)$ and $f_{YQ}(y)$ denote the distributions of Y scores in populations P and Q, respectively. Note, however, that because under a NEAT design test form X is only administered to population P and test form Y is only administered to population Q, the score distributions $f_{XQ}(x)$ and $f_{YP}(y)$ cannot be estimated from the collected data.[1] A common solution to this problem is to administer an anchor test, A, to both populations P and Q. The unobserved distributions can be estimated after imposing certain assumptions and marginalizing the resulting conditional score distributions $f_{XP}(x \mid a)$ and $f_{YQ}(y \mid a)$ over $f_{AT}(a)$, which is the (marginal) anchor scores distribution on T. Details of this derivation can be found in Sect. B.4.

The equating methods that will be described in Sects. 3.4.1 and 3.4.2 for the NEAT design make use of different assumptions in order to estimate either the unobserved distributions $f_{XQ}(x)$ and $f_{YP}(y)$ (i.e., for equipercentile transformations) or their corresponding location-scale parameters μ_{XQ}, σ^2_{XQ}, μ_{YP}, and σ^2_{YP} (i.e. for linear transformations).

[1]Technically, the score probability distributions are not identifiable (see, San Martín and González 2017).

3.3 Traditional Equating Methods for the EG, SG and CB Designs

In the previous section it was noted that no particular assumptions are needed when using the EG, SG, or CB designs. This means that we can estimate φ in definitions (3.1), (3.2) and (3.3) directly from the data. Thus, the means and the standard deviations for the linear equating in Eq. (3.2), or the means for mean equating in Eq. (3.3), can be estimated directly using the data at hand and then plugging the obtained quantities into their definitions of φ. Likewise, the equipercentile equating transformation in Eq. (3.1) can be estimated from the data without resorting to any particular assumptions. Examples of using **R** with the EG and SG designs are provided later in this chapter.

3.4 Traditional Equating Methods for the NEAT Design

Two common approaches that can be used either with linear or equipercentile equating transformations under a NEAT design are *chained equating* and *frequency estimation equating*. The first approach equates X to Y through A and does not make use of the concept of a synthetic population to obtain score distributions defined on T. To understand this better, let $\varphi_A(x)$ be the transformation that links scores in X to those in A and $\varphi_Y(a)$ be the transformation that links scores on A to Y. Chained equating produces the transformation from X to Y by using

$$\varphi(x) = \varphi_Y(x) = \varphi_Y(\varphi_A(x)). \tag{3.5}$$

Because $\varphi_A(x)$ is estimated on P and $\varphi_Y(a)$ is estimated on Q it is uncertain in which population the equating transformation $\varphi_Y(x)$ is valid. As will be seen later, some assumptions are needed for $\varphi_Y(x)$ to be a valid equating transformation defined on a common population.

The frequency estimation approach, on the other hand, makes use of the synthetic population to obtain the score distributions of X and Y in T from which the equating transformation is built. As we have seen in Sect. 3.2.2, under this approach not all of the score distributions involved are estimable from the data and some assumptions will be needed to obtain the equating transformation defined on T. These assumptions will be described in the sections where frequency estimation methods are introduced.

3.4.1 Linear Equating Methods for the NEAT Design

To perform a linear equating using the transformation in Eq. (3.2) under the NEAT design, we first need the means and standard deviations for the score distributions of X and Y defined on the target population T. Using the definitions in Eqs. (B.14) and (B.15), it can be shown (Brennan 2006) that the parameters for the synthetic population are defined as

$$\mu_{XT} = w_P\mu_{XP} + w_Q\mu_{XQ}, \tag{3.6}$$

$$\mu_{YT} = w_P\mu_{YP} + w_Q\mu_{YQ}, \tag{3.7}$$

$$\sigma^2_{XT} = w_P\sigma^2_{XP} + w_Q\sigma^2_{XQ} + w_Pw_Q[\mu_{XP} - \mu_{XQ}]^2, \tag{3.8}$$

and

$$\sigma^2_{YT} = w_P\sigma^2_{YP} + w_Q\sigma^2_{YQ} + w_Pw_Q[\mu_{YP} - \mu_{YQ}]^2. \tag{3.9}$$

As pointed out before, not all parameters are directly estimable from the information at hand. For instance, μ_{XQ} cannot be directly estimated because population Q is only administered test form Y. Likewise, σ^2_{YP} cannot be estimated because only test form X is administered in population P. The methods that are described next make different assumptions to obtain the parameters that are not directly estimable.

3.4.1.1 Tucker Equating

Tucker equating (Gulliksen 1950, pp. 299–300) relies on the following two assumptions: (i) the conditional expectations of X given A and Y given A are the same in both populations, i.e., $E_P(X \mid A) = E_Q(X \mid A)$ and $E_P(Y \mid A) = E_Q(Y \mid A)$; and (ii) the conditional variances of X given A and Y given A are the same in both populations, i.e., $Var_P(X \mid A) = Var_Q(X \mid A)$ and $Var_P(Y \mid A) = Var_Q(Y \mid A)$. Traditionally, the conditional expectations are assumed to be linear and the conditional variances are assumed to be constant, which is similar to the assumptions made in linear regression models. Thus, assumptions i and ii above can be equivalently reformulated by forcing slopes, intercepts and variances to be the same for both linear regressions (i.e., X regressed on A and Y regressed on A). Formally, if δ and γ are the intercept and slope parameters of the regressions in P and Q defined as

$$\delta_P = \mu_{XP} - \gamma_P\mu_{AP}, \quad \delta_Q = \mu_{XQ} - \gamma_Q\mu_{AQ}, \quad \gamma_P = \frac{\sigma_{XP,AP}}{\sigma^2_{AP}}, \quad \text{and} \quad \gamma_Q = \frac{\sigma_{YQ,AQ}}{\sigma^2_{AQ}}, \tag{3.10}$$

with $\sigma_{Z,W}$ representing the covariance between any two random variables Z and W, then by making $\gamma_P = \gamma_Q$ and $\delta_P = \delta_Q$ it can be shown that the means and variances of the synthetic target population are

$$\mu_{XT} = \mu_{XP} - w_Q\gamma_P[\mu_{AP} - \mu_{AQ}], \tag{3.11}$$

$$\mu_{YT} = \mu_{YQ} + w_P\gamma_Q[\mu_{AP} - \mu_{AQ}], \tag{3.12}$$

$$\sigma^2_{XT} = \sigma^2_{XP} - w_Q\gamma^2_P[\sigma^2_{AP} - \sigma^2_{AQ}] + w_Pw_Q\gamma^2_P[\mu_{AP} - \mu_{AQ}]^2, \tag{3.13}$$

and

$$\sigma^2_{YT} = \sigma^2_{YQ} + w_P\gamma^2_Q[\sigma^2_{AP} - \sigma^2_{AQ}] + w_Pw_Q\gamma^2_Q[\mu_{AP} - \mu_{AQ}]^2, \tag{3.14}$$

where γ_P is the slope for the regression of X on A in population P and γ_Q is the slope of the regression of Y on A in population Q.

To obtain the Tucker equating transformation, first plug γ_P and γ_Q defined in Eq. (3.10) into Eqs. (3.11), (3.12), (3.13) and (3.14) and then plug the latter into the general linear equating transformation in Eq. (3.2).

3.4.1.2 Nominal Weights Equating

A simplified version of Tucker equating is nominal weights equating which is typically used for small samples (Babcock et al. 2012; Albano 2016). It relies on the assumptions that test forms X and Y and the anchor test form A correlate perfectly within the populations P and Q and that the test forms have similar statistical properties. If J_x, J_y and J_a are the number of items in X, Y, and A, respectively, the following ratios are useful approximations of the γ terms in Eq. (3.10)

$$\gamma_P = \frac{J_x}{J_a} \text{ and } \gamma_Q = \frac{J_y}{J_a}. \tag{3.15}$$

To obtain the equating transformation using nominal weights, proceed as with Tucker equating using Eq. (3.15) instead of Eq. (3.10).

3.4.1.3 Levine Observed-Score Equating

Levine observed-score equating was first proposed by Levine (1955). In what follows we adopt the description of the Levine observed-score method given in Kolen and Brennan (2014). The assumptions in Levine observed-score equating are based on the classical test theory model (see, e.g., Lord and Novick 1968; Crocker and Algina 1986). The idea is that for each of the involved test forms the observed scores X, Y and A can be decomposed as the sum of true scores τ_X, τ_Y, and τ_A and errors E_X, E_Y, and E_A in the following way:

$$X = \tau_X + E_X, \tag{3.16}$$

$$Y = \tau_Y + E_Y, \tag{3.17}$$

and

$$A = \tau_A + E_A. \tag{3.18}$$

In Levine observed-score equating, the three assumptions are stated in terms of true scores even though observed test scores are used in practice for equating. These assumptions are:

1. The correlation between the true scores on test forms X and Y with the anchor test form A is 1.
2. The coefficients of a linear regression of the true scores on test forms X and Y on A are the same.
3. The measurement error variance across populations is the same for test forms X, Y and A.

Using these assumptions it is possible to show that γ slopes are defined in terms of true-score standard deviations, which in practice are not observable. Although it is still possible to obtain estimates for the γ terms by following the assumptions of classical test theory, the results used here and shown below are derived from what is referred to as the *classical congeneric model* (see, Kolen and Brennan 2014, Sects. 4.2.5 and 4.2.6)

$$\gamma_P = \frac{\sigma^2_{XP}}{\sigma_{XP,AP}}, \quad \text{and} \quad \gamma_Q = \frac{\sigma^2_{YQ}}{\sigma_{YQ,AQ}}. \tag{3.19}$$

To obtain the Levine observed-score linear transformation, the values in Eq. (3.19) are plugged into Eqs. (3.11), (3.12), (3.13) and (3.14). The final step is to plug the resulting means and variances into the general linear equating transformation in Eq. (3.2).

Note that we have defined Levine observed-score equating by resorting on the classical congeneric model and assuming that an internal anchor is used. The definitions for Levine observed-score equating with an external anchor can be found for example, in Kolen and Brennan (2014, p. 115).

3.4.1.4 Levine True-Score Equating

A method that relies on the same assumptions as the Levine observed-score method is the Levine true-score method (Levine 1955). The conceptual difference is that instead of observed scores, true scores are equated.

After some algebra and using the assumptions from classical test theory, the linear equating function for Levine true-score equating becomes

$$\varphi(\tau_X; \mu_{XP}, \mu_{YQ}, \mu_{AP}, \mu_{AQ}, \gamma_P, \gamma_Q) = \frac{\gamma_Q}{\gamma_P}[\tau_X - \mu_{XP}] + \mu_{YQ} + \gamma_Q[\mu_{AP} - \mu_{AQ}], \tag{3.20}$$

where γ_Q and γ_P are defined in Eq. (3.19). For a theoretical justification and for more information about the Levine true-score equating refer to Kolen and Brennan (2014), Hanson (1991), or Hanson et al. (1993).

3.4.1.5 Chained Linear Equating

Chained linear equating (Angoff 1971) can be performed by calculating the means and standard deviations of the test scores and anchor test scores within the populations P and Q and then composing the estimated linear transformations that links X to A with the estimated linear transformation linking A to Y (see Eq. (3.5)). Chained linear equating results in the following equating transformation

$$\varphi(x) = \mu_{YQ} + \frac{\sigma_{YQ}}{\sigma_{AQ}}[\mu_{AP} - \mu_{AQ}] - \frac{\sigma_{YQ}/\sigma_{AQ}}{\sigma_{XP}/\sigma_{AP}}\mu_{XP} + \frac{\sigma_{YQ}/\sigma_{AQ}}{\sigma_{XP}/\sigma_{AP}}x. \tag{3.21}$$

3.4.2 Equipercentile Equating Methods for the NEAT Design

As previously mentioned, there are two possible approaches to perform equipercentile equating under a NEAT design: frequency estimation and chained equating. Both of these approaches are described next.

3.4.2.1 Frequency Estimation

The frequency estimation equating method (Angoff 1971; Petersen et al. 1989; Braun and Holland 1982) makes use of an anchor test to estimate both the X and Y score distributions in the target population T. From the obtained score distributions, percentile ranks are obtained and the forms are equated using equipercentile equating.

Formally, the conditional distribution of test scores X given an anchor test score A defined on population P is denoted as $f_{XP}(x|a)$ (see Sect. 3.2.2). Similarly for Y, the conditional distribution defined on population P is denoted as $f_{YP}(y|a)$. The corresponding conditional distributions defined on population Q are denoted as $f_{XQ}(x|a)$ and $f_{YQ}(y|a)$, respectively. The marginal distributions of anchor test scores are accordingly defined as $f_{AP}(a)$ and $f_{AQ}(a)$ for both populations. In order to obtain the (marginal) score distributions defined for T, the following assumptions are made

1. $f_{XP}(x|a) = f_{XQ}(x|a)$, i.e., the conditional X score distributions are the same for any target population T of the form given in Eq. (3.4).
2. $f_{YP}(y|a) = f_{YQ}(y|a)$, i.e., the conditional Y score distributions are the same for any target population T of the form given in Eq. (3.4).

With these assumptions, it follows from Eqs. (B.14) and (B.15) that

$$f_{XT}(x) = w_P f_{XP}(x) + w_Q \sum_a f_{XP}(x \mid a) f_{AQ}(a), \tag{3.22}$$

and

$$f_{YT}(y) = w_Q f_{YQ}(y) + w_P \sum_a f_{YQ}(y \mid a) f_{AP}(a). \tag{3.23}$$

3.4.2.2 Chained Equipercentile Equating

In chained equipercentile equating (Dorans 1990; Livingston et al. 1990), the test scores X and Y are connected or "chained" together through the anchor test as explained in Sect. 3.4. First, test form X is equated to test form A and then test form A is equated to test form Y. The chained equipercentile equating transformation can formally be defined as

$$\varphi_Y(x) = F_{YQ}^{-1}(F_{AQ}(F_{AP}^{-1}(F_{XP}(x)))) = \varphi_{YQ}(\varphi_{AP}(x)). \tag{3.24}$$

As mentioned in Sect. 3.4, some assumptions have to be made to ensure that $\varphi_Y(x)$ is properly defined on T and these are the following

1. $\varphi_{AP}(x) = \varphi_{AQ}(x) = \varphi_A(x)$, i.e., the equating from test form X to the anchor test form A for the target population T is the same for any T of the form given in Eq. (3.4).
2. $\varphi_{YP}(a) = \varphi_{YQ}(a) = \varphi_Y(a)$, i.e., the equating from anchor test form A to test form Y for the target population T is the same for any T of the form given in Eq. (3.4).

Under these assumptions, chained equating can be shown to produce a valid equating transformation defined for a common population T.

3.4.2.3 Braun-Holland Equating

The Braun-Holland equating method (Braun and Holland 1982) is a *hybrid* between linear and equipercentile methods for the NEAT design. This method is a linear version of the frequency estimation equating method, which is described in

Sect. 3.4.2.1. It is based on estimates of the means and standard deviations of the synthetic target distributions $f_{XT}(x)$ and $f_{YT}(y)$ given in Eq. (3.22). The mean and standard deviation in the synthetic target distributions for X are defined as

$$\mu_{XT} = \sum_x x f_{XT}(x), \tag{3.25}$$

$$\sigma^2_{XT} = \sum_x [x - \mu_{XT}]^2 f_{XT}(x),$$

and for Y are defined as

$$\mu_{YT} = \sum_y y f_{YT}(y), \tag{3.26}$$

$$\sigma^2_{YT} = \sum_y [y - \mu_{YT}]^2 f_{YT}(y).$$

Parameters in Eqs. (3.25) and (3.26) are then plugged into Eq. (3.2) to obtain the Braun-Holland linear equating transformation.

3.5 Examples with the `equate` Function

In this section, different examples of the previously described traditional equating methods are illustrated using the `equate()` function in the **equate** package. To get a deeper understanding, we start with a short description of `equate()` that includes the general function call. Note that although we have discussed four data collection designs (EG, SG, CB, and NEAT), only three of them (EG, SG, and NEAT) can currently be implemented directly in the **equate** package. Implementing equating under the CB design is still possible, but it requires additional steps depending on the way the score data are used (see Sect. 1.3.1.3).

3.5.1 The `equate` Function

The `equate()` function implements all of the traditional equating methods we have introduced so far in this chapter. The general function call is

```
equate(x, y, type = c("identity", "mean", "linear",
"general linear", "circle-arc", "equipercentile"),
method = c("none","nominal weights","tucker","levine",
"frequency estimation", "chained", "braun/holland"),
name, lowp, highp, boot = FALSE, verbose = TRUE, ...)
```

The arguments x and y are used to provide X and Y scores as a frequency table of class "freqtab" (see Sect. 2.2). If y is not provided, then an SG design is assumed. The argument type specifies the kind of equating transformation to be used (e.g., equipercentile, linear, or mean as defined in Eqs. (3.1), (3.2), and (3.3), respectively). Other types of equating transformation not described so far (i.e., identity, general linear, and circle-arc) will be discussed in Sect. 3.6. The argument method is used to specify which equating method is to be performed, especially for the case when score data have been collected under the NEAT design. Possible methods include all of those discussed in this chapter for the NEAT design (nominal weights, Tucker, Levine observed and true score, frequency estimation, chained equating, and Braun-Holland). The default method "none" is used to perform an equating under the EG or the SG design. The optional arguments lowp (or highp) allow the user to set the lowest (or the highest) score expected by chance. Using the argument boot=TRUE allows the user to obtain bootstrap standard errors (see Sect. 3.8). Finally, the argument verbose is used to decide between two display options for the output. If verbose=FALSE, then only the equated values are displayed, whereas if verbose=TRUE, which is the default, then a more complete output including relevant descriptive statistics for the scores will be displayed.

Note, **equate** allows the use of short names when writing the code, e.g. "equip" and "e" can be used instead of "equipercentile", "m" is used for "mean", and "l" is used for "linear". Similarly, "frequency estimation" can also be written "freq" or simply "f". Likewise, "t" and "b" can be used for Tucker and Braun-Holland equating, respectively.

3.5.2 Examples Under the EG and SG Designs

To illustrate the use of the equate() function under the EG design, we start with a simple example where equipercentile equating is performed using the ACTmath data described in Chap. 2. Although the necessary steps for using the ACTmath data as score frequency distributions were already described in Sect. 2.2.2, we repeat the needed code here for completeness:

```
> library(equate)
> xACT <- as.freqtab(ACTmath[,c(1,2)])
> yACT <- as.freqtab(ACTmath[,c(1,3)])
```

Once score frequency distributions have been obtained, equipercentile equating can be performed using the following code:

```
> eq.equipercentile <- equate(xACT, yACT,
+ method = "none",type = "equipercentile")
```

Because under the EG design no special assumptions are needed to estimate the equating transformation (see Sect. 3.2.1), the argument method is set to "none". Typing eq.equipercentile gives the following output:

```
> eq.equipercentile

Equipercentile Equating: xACT to yACT

Design: equivalent groups

Smoothing Method: none

Summary Statistics:
    mean   sd skew kurt  min  max    n
x  19.85 8.21 0.38 2.30 1.00 40.0 4329
y  18.98 8.94 0.35 2.15 1.00 40.0 4152
yx 18.98 8.94 0.35 2.15 0.98 39.9 4329
```

The output indicates that an equipercentile equating from X to Y was performed under the EG design and that no smoothing method was applied to the score distributions. The summary statistics indicate the mean, standard deviation, skewness, kurtosis, minimum, maximum and the sample sizes for the raw scores (with row labels `x` and `y`) and the equated scores (with row label `yx`). Note that the mean, standard deviation, skewness and kurtosis (the first four moments of the score distribution) are the same for *Y* and for the equated values. This happens because the equipercentile equating transformation is built from the whole score distributions in comparison with the linear and mean equating transformation where, as we have seen in Sect. 3.1, only the first two moments of the score distribution play a role in the equating. In the examples that come below this will be explicitly illustrated.

The displayed output for an `equate` object contains only some information. Such object could contains further useful information. If we want to see the complete list of output elements that are available, the `str()` function can be applied to an equating object. For example, applying the `str()` function to the `eq.equipercentile` object gives the following output:

```
> str(eq.equipercentile)
List of 9
 $ name        : chr "Equipercentile Equating: xACT to yACT"
 $ type        : chr "equipercentile"
 $ method      : chr "none"
 $ design      : chr "equivalent groups"
 $ x           : freqtab [1:41(1d)] 0 1 1 3 9 18 59 67 91 144 ...
  ..- attr(*, "dimnames")=List of 1
  .. ..$ total: chr [1:41] "0" "1" "2" "3" ...
  ..- attr(*, "class")= chr [1:2] "freqtab" "table"
  ..- attr(*, "design")= chr "eg"
 $ y           : freqtab [1:41(1d)] 0 1 3 13 42 59 95
                 131 158 161 ...
  ..- attr(*, "dimnames")=List of 1
  .. ..$ total: chr [1:41] "0" "1" "2" "3" ...
  ..- attr(*, "class")= chr [1:2] "freqtab" "table"
  ..- attr(*, "design")= chr "eg"
 $ concordance :'data.frame':   41 obs. of  3 variables:
  ..$ scale: num [1:41] 0 1 2 3 4 5 6 7 8 9 ...
  ..$ yx   : num [1:41] -0.5 0.98 1.65 2.29 2.89 ...
```

```
  ..$ se    : num [1:41] 0 0.831 0.521 0.821 0.295 ...
 $ points        :'data.frame':    2 obs. of  2 variables:
  ..$ low : num [1:2] 0 0
  ..$ high: num [1:2] 40 40
 $ smoothmethod: chr "none"
 - attr(*, "class")= chr "equate"
```

The output shows that the eq.equipercentile object is a list of nine elements. Each of the elements can be retrieved by typing the object's name and writing as a suffix the dollar sign ($) followed by any of the available options. For instance, if the interest is in retrieving the equated values, then the concordance (or short coded here as con or conc) element in the list must be selected

```
> eq.equipercentile$con
   scale         yx        se
1      0 -0.5000000 0.0000000
2      1  0.9795564 0.8305507
3      2  1.6462231 0.5210048
4      3  2.2856318 0.8209735
5      4  2.8931979 0.2950250
6      5  3.6204666 0.1478062
7      6  4.4996535 0.2541079
8      7  5.5148375 0.1581823
9      8  6.3124157 0.1969085
10     9  7.2242386 0.1761150
11    10  8.1606665 0.1731162
12    11  9.1826961 0.1951590
13    12 10.1858956 0.1799526
14    13 11.2513015 0.2310924
15    14 12.3896334 0.2431244
16    15 13.3928909 0.2138483
17    16 14.5240050 0.2763547
18    17 15.7169010 0.2617253
19    18 16.8234423 0.3383500
20    19 18.0092239 0.2826068
21    20 19.1647208 0.2947267
22    21 20.3676007 0.3298667
23    22 21.4556277 0.3182666
24    23 22.6871228 0.3864620
25    24 23.9156570 0.3554621
26    25 25.0291585 0.3013309
27    26 26.1612293 0.3683138
28    27 27.2632870 0.3532292
29    28 28.1800647 0.3069083
30    29 29.1424331 0.3422039
31    30 30.1304817 0.2896340
32    31 31.1297014 0.3267974
33    32 32.1357069 0.3309308
34    33 33.0780678 0.3047704
35    34 34.0171864 0.3079831
36    35 35.1016041 0.3043501
37    36 36.2425502 0.3239977
38    37 37.1247622 0.2713679
```

```
39     38 38.1320883 0.3430113
40     39 39.0807346 0.2017926
41     40 39.9005544 0.2787227
```

This element contains the score scale, the equated values, and standard errors.[2] If we only want to obtain the equated values, we add an extra suffix term (`yx`) as shown in the following code (output not shown):

```
> eq.equipercentile$conc$yx
```

The output stored in an equating object and displayed by the `str()` function will depend on the tasks that have been performed by the `equate()` function. This means that if, for instance, the `boot` argument was added, then bootstrap standard errors will also be available from the created equating object. Bootstrap standard errors will be described in Sect. 3.8.

Using the previously created score frequency distributions `xACT` and `yACT`, we can easily perform a mean equating under the EG design as follows

```
> eq.mean <- equate(xACT, yACT, type = "m")
```

Note that this time we have not typed the argument `method` as the default value is `"none"`. Also, instead of writing `"mean"` in the `type` argument, the shorthand `"m"` was used. The displayed output for mean equating differs from that for the equipercentile equating shown above and it adds the elements `intercept`, `slope`, `cx`, `cy`, `sx`, and `sy`.

```
> eq.mean

Mean Equating: xACT to yACT

Design: equivalent groups

Summary Statistics:
    mean   sd skew kurt  min   max    n
x  19.85 8.21 0.38 2.30 1.00 40.00 4329
y  18.98 8.94 0.35 2.15 1.00 40.00 4152
yx 18.98 8.21 0.38 2.30 0.13 39.13 4329

Coefficients:
intercept    slope         cx         cy         sx         sy
  -0.8726   1.0000    20.0000    20.0000    40.0000    40.0000
```

The first two elements, the `intercept` and the `slope`, correspond to the intercept, $\mu_Y - \frac{\sigma_Y}{\sigma_X}\mu_X$, and the slope, $\frac{\sigma_Y}{\sigma_X}$, of the linear equating function in Eq. (3.2). Because mean equating is being performed, the slope is equal to one. The last four elements `cx`, `cy`, `sx`, and `sy` are parameters used in the general linear equating approach introduced in Albano (2015).

[2] Analytic standard errors are displayed when available.

Note also that this time only the mean of the score values *Y* coincides with that of the equated values. As pointed out before, this is the case because in mean equating only the first moment of the score distribution plays a role in the equating. The reader can verify that if a linear equating is performed, then only the mean and standard deviation of the *Y* scores and equated scores will coincide. Linear equating under the EG design can easily be obtained by changing the type of the equating in the code to `"linear"` (or equivalently `"l"` as a shorthand) as follows:

```
> eq.linear <- equate(xACT, yACT, type = "l")
```

Performing equating under the SG design only requires a few modifications to the code presented above. Indeed, the main modification is that data formatted according to a SG design should be used as an input (see Sect. 2.2.3). We now use the `Math20SG` data to illustrate mean, linear, and equipercentile equating under the SG design. The `sg.data` object previously created in Sect. 2.2.3 is already formatted as score frequency distributions and can be used as input for linear equating under the SG design:

```
> sg.linear <- equate(sg.data, type = "l")
> sg.linear

Linear Equating: sg.data to sg.data

Design: single group

Summary Statistics:
    mean   sd  skew kurt  min   max    n
x  10.82 3.81  0.00 2.53 0.00 20.00 1453
y  10.39 3.59 -0.01 2.48 2.00 19.00 1453
yx 10.39 3.59  0.00 2.53 0.19 19.04 1453

Coefficients:
intercept   slope       cx       cy       sx       sy
   0.1937  0.9424  10.0000  10.0000  20.0000  20.0000
```

Similarly as for the EG design, we can perform both mean equating and equipercentile equating under the SG design by simply modifying the method argument as `method="mean"` and `method="equipercentile"`, respectively.

In the following example, the three equating methods implemented above under the SG design are compared graphically. The `plot()` function applied to an `equate` object will by default produce a plot of the corresponding equating transformation. When applied to several equating objects, it will draw all the equating functions that are part of passed equating objects. First, we use the following code to obtain mean, linear, and equipercentile equating under the SG design:

```
> mod.sg1 <- equate(sg.data,type="equipercentile")
> mod.sg2 <- equate(sg.data,type="mean")
> mod.sg3 <- equate(sg.data,type="linear")
```

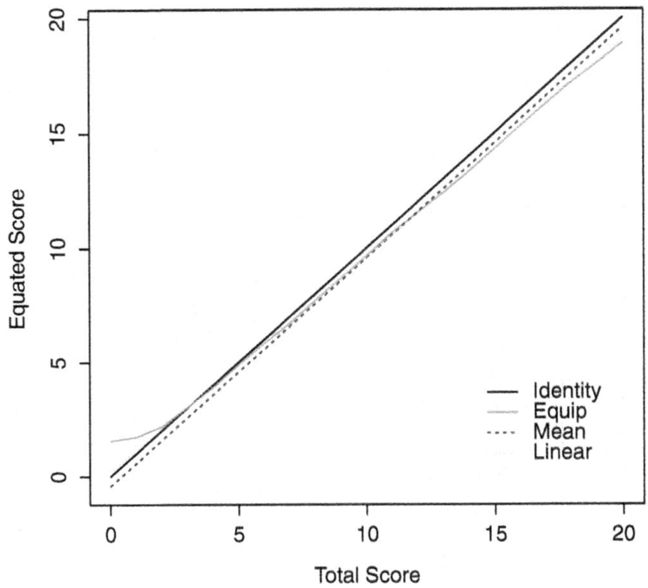

Fig. 3.1 The identity, equipercentile, mean, and linear equating transformations

Figure 3.1 shows the obtained equating transformations and was produced using the following code:

```
> plot(mod.sg1,mod.sg2,mod.sg3,lty=c(1,2,3,4),
+ col=c(3,4,5,6))
```

Note that the identity equating transformation $\varphi(x) = x$ is plotted by default but can be omitted by adding the argument `addident = FALSE` to the `plot()` call.

As a final example to illustrate traditional equating methods under the EG design, we give the code to reproduce Table 2.7 in Kolen and Brennan (2014), where a comparison between mean, linear, and equipercentile equating is made using the `ACTmath` data. Using the previously created objects `xACT` and `yACT`, the different equating transformations are obtained using the argument `type` with values `"m"`, `"l"`, and `"e"` meaning that mean, linear, and equipercentile equating are being performed.

```
> eq.mean   <- equate(xACT, yACT, type = "m")
> eq.linear <- equate(xACT, yACT, type = "l")
> eq.equipercentile <- equate(xACT, yACT, type = "e")
```

We use the equated values stored in each of the created equating objects to mimic Table 2.7 in Kolen and Brennan (2014).

```
> Table2.7 <- cbind(0:40,eq.mean$con$yx,
+ eq.linear$con$yx, eq.equipercentile$con$yx)
> colnames(Table2.7) <- c("Score", "Mean", "Linear",
```

```
+ "Equipercentile")
> Table2.7
       Score       Mean     Linear Equipercentile
 [1,]     0 -0.8726221 -2.6319708     -0.5000000
 [2,]     1  0.1273779 -1.5433493      0.9795564
 [3,]     2  1.1273779 -0.4547278      1.6462231
 [4,]     3  2.1273779  0.6338937      2.2856318
 [5,]     4  3.1273779  1.7225152      2.8931979
 [6,]     5  4.1273779  2.8111367      3.6204666
 [7,]     6  5.1273779  3.8997582      4.4996535
 [8,]     7  6.1273779  4.9883797      5.5148375
 [9,]     8  7.1273779  6.0770012      6.3124157
[10,]     9  8.1273779  7.1656227      7.2242386
[11,]    10  9.1273779  8.2542443      8.1606665
[12,]    11 10.1273779  9.3428658      9.1826961
[13,]    12 11.1273779 10.4314873     10.1858956
[14,]    13 12.1273779 11.5201088     11.2513015
[15,]    14 13.1273779 12.6087303     12.3896334
[16,]    15 14.1273779 13.6973518     13.3928909
[17,]    16 15.1273779 14.7859733     14.5240050
[18,]    17 16.1273779 15.8745948     15.7169010
[19,]    18 17.1273779 16.9632163     16.8234423
[20,]    19 18.1273779 18.0518378     18.0092239
[21,]    20 19.1273779 19.1404593     19.1647208
[22,]    21 20.1273779 20.2290808     20.3676007
[23,]    22 21.1273779 21.3177023     21.4556277
[24,]    23 22.1273779 22.4063238     22.6871228
[25,]    24 23.1273779 23.4949453     23.9156570
[26,]    25 24.1273779 24.5835668     25.0291585
[27,]    26 25.1273779 25.6721883     26.1612293
[28,]    27 26.1273779 26.7608098     27.2632870
[29,]    28 27.1273779 27.8494313     28.1800647
[30,]    29 28.1273779 28.9380528     29.1424331
[31,]    30 29.1273779 30.0266743     30.1304817
[32,]    31 30.1273779 31.1152958     31.1297014
[33,]    32 31.1273779 32.2039173     32.1357069
[34,]    33 32.1273779 33.2925388     33.0780678
[35,]    34 33.1273779 34.3811603     34.0171864
[36,]    35 34.1273779 35.4697818     35.1016041
[37,]    36 35.1273779 36.5584033     36.2425502
[38,]    37 36.1273779 37.6470248     37.1247622
[39,]    38 37.1273779 38.7356463     38.1320883
[40,]    39 38.1273779 39.8242678     39.0807346
[41,]    40 39.1273779 40.9128893     39.9005544
```

3.5.3 *Examples Under the NEAT Design*

3.5.3.1 Linear Methods

We start by comparing four linear equating methods under the NEAT design: Tucker, chained, Levine observed-score, and Levine true-score equating. The first two lines of code below read in the data, and the subsequent lines perform the four different equating methods.

```
> nex <- freqtab(KBneat$x, scales = list(0:36, 0:12))
> ney <- freqtab(KBneat$y, scales = list(0:36, 0:12))
> neq1 <-equate(nex,ney,type="linear",method="tuck",
+ ws=1)
> neq2 <-equate(nex,ney,type="linear",method="chain")
> neq3 <-equate(nex,ney,type="linear",method="levine",
+ ws=1)
> neq4 <-equate(nex,ney,type="linear",method="levine",
+ lts=TRUE,ws=1)
```

In all cases, the argument `type` is set to `"linear"` because a linear transformation such as the one shown in Eq. (3.2) is used for the equating. However, different methods are specified because all of them make different assumptions for obtaining the location and scale parameters needed to estimate $\varphi(x)$. In the case of the Tucker and the Levine methods, setting the argument `ws=1` means that the weight w_P in Eq. (3.4) is set to 1 (see also Eqs. (3.11), (3.12), (3.13) and (3.14)). For the Levine method, the default method is observed-score equating so to obtain the Levine true-score equating the additional argument `lts=TRUE` has been added in `neq4`. For the four methods, the displayed output will be very similar to the output described earlier for the `eq.mean` object, except for the chained equating in object `neq2` where the elements `cx`, `cy`, `sx`, and `sy` are not displayed.

The obtained equated values can be used to reproduce Table 4.5 of Kolen and Brennan (2014) using the following code:

```
> compneq<-round(cbind(xscale=0:36,tucker=neq1$conc$yx,
+ chain=neq2$conc$yx,levineOS=neq3$conc$yx,
+ levineTS=neq4$conc$yx),4)
> Table45 <- compneq[c(1,11,21,31,37),]
> Table45
     xscale  tucker   chain levineOS levineTS
[1,]      0  0.5368  0.3937   0.2513   0.2912
[2,]     10 10.8263 10.6064  10.3630  10.3777
[3,]     20 21.1157 20.8191  20.4747  20.4641
[4,]     30 31.4052 31.0318  30.5863  30.5506
[5,]     36 37.5789 37.1595  36.6533  36.6024
```

3.5.3.2 Equipercentile Methods

To illustrate how to perform both equipercentile frequency estimation and chained equipercentile equating, we use data from Kolen and Brennan (2014) as described

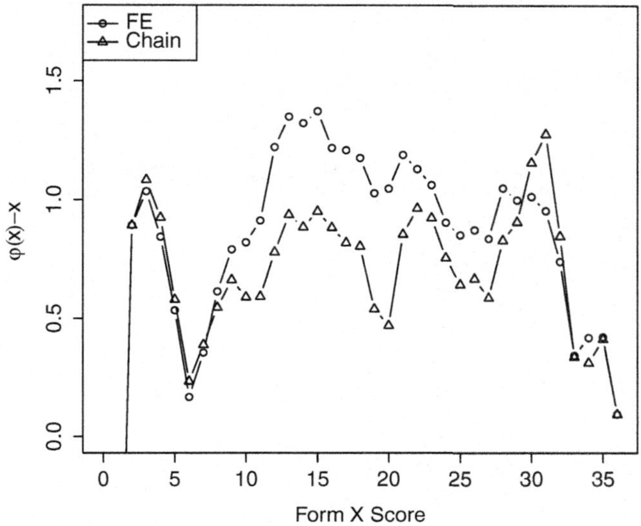

Fig. 3.2 Relationship between frequency estimation (FE) and chained equating

in Chap. 2. The data set is also available in the **equate** package and it is stored with the name `KBneat`. The following code performs both the frequency estimation equating and chained equipercentile equating:

```
> neq5 <- equate(nex, ney, type = "equipercentile",
+ method = "freq", ws = 1)
> neq6 <- equate(nex, ney, type = "equipercentile",
+ method = "chain")
```

The argument `ws` for the frequency estimation method corresponds to the w_P weight used in the definition of synthetic population in Eq. (3.4), and in this case it is set to 1, meaning that $w_Q = 0$.

The following code reproduces part of Figure 5.4 in Kolen and Brennan (2014). Because the modified frequency estimation method is not implemented in **equate**, only two curves are shown in Fig. 3.2.

```
> plot(0:36,neq5$con$yx-0:36,type="b",pch=1,ylim=c(0.00,
+ 1.75),ylab=expression(varphi(x)-x),xlab="Form X Score")
> lines(0:36,neq6$con$yx-0:36,type="b",pch=2)
> legend("topleft",c("FE","Chain"),lty=c(1,1),pch=c(1,2))
```

3.5.3.3 Comparison Between Linear and Equipercentile Methods

Kolen and Brennan (2014, Section 5.1.6) make a comparison between equipercentile frequency estimation, Tucker, Levine observed-score, and Braun-Holland equating methods. In the case of equipercentile frequency estimation,

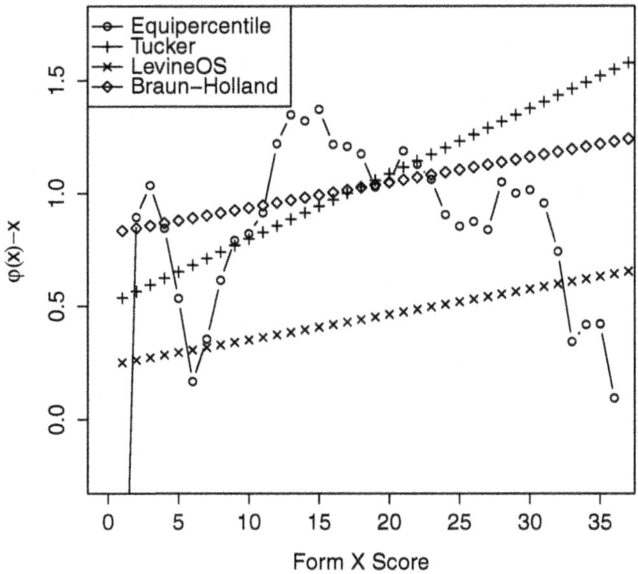

Fig. 3.3 The relationship between equipercentile frequency estimation and the Tucker, Levine observed-score, and Braun-Holland linear equating methods

both smoothed (using cubic splines postsmoothing) and unsmoothed results are considered. The code shown below can be used to reproduce Figure 5.3 in Kolen and Brennan (2014). Note that we first obtain the results for the Braun-Holland linear equating and store them in the object `neq7`. Also, note that because **equate** does not implement postsmoothing, one of the curves (as well as the standard error bands) in the original figure is omitted in Fig. 3.3.

```
> neq7 <- equate(nex,ney,type="linear",
+ method="braun/holland",ws=1)
> plot(0:36,neq5$con$yx-0:36,type="b",pch=1, ylim=c(-0.25,
+ 1.75),ylab=expression(varphi(x)-x),xlab="Form X Score")
> lines(neq1$con$yx-0:36,type="b",pch=3)
> lines(neq3$con$yx-0:36,type="b",pch=4)
> lines(neq7$con$yx-0:36,type="b",pch=5)
> legend("topleft",c("Equipercentile","Tucker","LevineOS",
+ "Braun-Holland"),lty=c(1,1,1,1),pch=c(1,3,4,5))
```

Log-linear smoothing in the analyses is easily incorporated in the code using the `smoothmethod` argument. For instance, under the NEAT design, the following code can be used to perform equipercentile frequency estimation equating using presmoothed distributions

```
> neq8 <- equate(nex,ney,type="eq",method ="freq",
+ smoothmethod="log",degrees=list(3,3), ws=1)
```

Note that the inclusion of the `smoothmethod` argument inside the `equate()` function is only implemented for equipercentile equating.

3.5.4 Examples Using the ADM Data Under the NEAT Design

In order to examine different equating methods for the NEAT design we use the admissions ADM data described in Chap. 2. For a NEAT design the previously defined **R** objects `neat.x1` and `neat.y1` are used. In what follows we show the necessary code to implement Tucker, frequency estimation, Braun-Holland, and chained equating using the ADM data. Because the obtained outputs are similar to the ones we have obtained in previous sections for other data sets, they are omitted.

```
> Aeq1 <-equate(neat.x1,neat.y1,type="linear", method =
+ "tuck", ws = 1)
> Beq2 <-equate(neat.x1,neat.y1,type="linear", method =
+ "braun", ws = 1)
> Ceq3 <-equate(neat.x1,neat.y1,type="equip", method =
+ "freq", ws = 1)
> Deq4 <-equate(neat.x1,neat.y1,type="equip", method =
+ "chain")
```

To compare the obtained equated values under the different equating methods, the following code is used to create a table of all equated values (only the first 15 records are shown in the output)

```
> compADM<-round(cbind(xscale = 0:80, Tucker =
+ Aeq1$conc$yx,Braun = Beq2$conc$yx,
+ FE = Ceq3$conc$yx, Chain = Deq4$conc$yx), 2)
> head(compADM,n=15)
      xscale Tucker Braun    FE Chain
 [1,]      0  -0.34 -0.16 -0.50 -0.50
 [2,]      1   0.69  0.86 -0.50 -0.50
 [3,]      2   1.71  1.89 -0.50 -0.50
 [4,]      3   2.74  2.91 -0.50 -0.50
 [5,]      4   3.77  3.93 -0.50 -0.50
 [6,]      5   4.79  4.96 -0.50 -0.50
 [7,]      6   5.82  5.98 -0.50 -0.50
 [8,]      7   6.84  7.00 -0.50 -0.50
 [9,]      8   7.87  8.03 -0.50 -0.50
[10,]      9   8.90  9.05  8.78 -0.50
[11,]     10   9.92 10.07  9.33  8.62
[12,]     11  10.95 11.10  9.66  8.75
[13,]     12  11.98 12.12  9.66  8.75
[14,]     13  13.00 13.14 10.09  8.88
[15,]     14  14.03 14.17 11.41  9.38
```

3.6 Additional Features in equate

There are also additional features in the **equate** package, including other equating methods not described so far, and the possibility to obtain the bootstrap standard error of equating. One feature is the use of a circle-arc choice of the equating transformation (Livingston and Kim 2009) for Tucker, Braun-Holland, Levine and chained equating. A general description of the circle-arc method is given in Albano (2016) together with an example.

Throughout this chapter we have considered the case of only one anchor test form. However, nothing prevents us from using multiple anchor test forms as described in Albano (2016). In **equate**, this option is valid for frequency estimation, Braun-Holland, Tucker, and nominal weights equating.

Bootstrap standard errors are a useful tool when comparing different equating methods and this will be illustrated later in this chapter.

3.7 Performing Traditional Equating Methods with SNSequate

Mean, linear and equipercentile equating can also be performed using the functions `mea.eq()`, `lin.eq()`, and `eqp.eq()`, respectively in **SNSequate**. A notable difference is that frequency tables are not needed in **SNSequate**, but instead the full vectors of total scores are directly used as inputs for equating. Another difference is that the term `concordance` in **equate** is replaced by the term `resu` to obtain the resulting equated values in **SNSequate**. Using the `ACTmath` data included in **SNSequate**, the following code reproduces the previous described Table 2.7 in Kolen and Brennan (2014):

```
> library(SNSequate)
> act.m <- mea.eq(rep(0:40, ACTmath[,2]), rep(0:40,
+ ACTmath[,3]),0:40)
> act.l <- lin.eq(rep(0:40, ACTmath[,2]), rep(0:40,
+ ACTmath[,3]),0:40)
> act.e <- eqp.eq(rep(0:40, ACTmath[,2]), rep(0:40,
+ ACTmath[,3]),0:40)
```

In each case, the first two arguments correspond to the whole vector of scores for tests forms X and Y, respectively. Note that because the score data in the `ACTmath` object are stored as frequencies, the `rep()` function helps in expanding score frequencies across the score scale so as to obtain a score vector with a length equal to the number of test takers that have been administered the tests. The third argument indicates the range of score values that are to be equated – in this case, the whole score scale. For the three functions, the output corresponds to the score scale and the corresponding equated values according to the specified range of values in the third argument.

A table showing equated scores using the three methods is obtained as follows:

```
> Table2.7s <- cbind(0:40, act.m$resu, act.l$resu,
+ act.e$resu)
+ colnames(Table2.7s) <- c("Score", "Mean", "Linear",
+ "Equipercentile")
> Table2.7s
```

The reader can verify that this table is actually identical to the one obtained using **equate** in Sect. 3.5.2 and to the one shown as Table 2.7 in Kolen and Brennan (2014).

3.8 Comparing Traditional Test Equating Methods

The performance of different test equating methods can be examined in different ways. As seen in Chap. 1, equating transformations are statistical estimators and as such they are subject to systematic and random errors. The systematic error in statistics is typically examined measuring bias, whereas random errors are examined with standard errors. In this chapter the bias, standard errors, and root mean squared errors (RMSE), are defined, and we describe how they can be obtained using the **equate** package.

3.8.1 Bootstrap Standard Errors of Equating

Suppose for a moment that a very large number of replications of an equating procedure can be obtained using samples from a population of examinees. This means that, for each score value, there is a large collection of equated values obtained using each sample in the replications. For each score value, the standard error of equating can be defined as the standard deviation of the collected equated values over the replications (Kolen and Brennan 2014). However, in real life practice only one sample of data is usually available, and thus methods to obtain the standard error of equating should use the score data at hand. One such method is the bootstrap method (Efron 1982; Efron and Tibshirani 1993) that can be used to obtain the standard error of any statistic of interest using a large number of random samples with replacement from the (unique) observed sample data. This method is extremely useful in the context of equating because for some equating transformations it is sometimes very difficult (if not impossible) to derive explicit formulas of the standard errors.

In the case of equating functions, the following algorithm can be used to obtain standard errors of equated values using bootstrap:

1. Draw a random sample of size n_X with replacement from the observed sample of n_X test takers that have been administered test form X.
2. Draw a random sample of size n_Y with replacement from the observed sample of n_Y test takers that have been administered test form Y.
3. Estimate the equating transformation using the data from the bootstrap samples obtained in step 1 and 2.
4. Replicate step 1 through 3 L times, which yields L bootstrap estimates of the equating transformation.
5. Estimate the standard error using

$$\mathrm{SE}(x_i) = \sqrt{\frac{1}{L}\sum_{l=1}^{L}[\hat{\varphi}_l(x_i) - \bar{\hat{\varphi}}_l(x_i)]^2} \tag{3.27}$$

where $\bar{\hat{\varphi}}_l(x_i) = \frac{1}{L}\sum_{l=1}^{L}\hat{\varphi}_l(x_i)$, and $\hat{\varphi}_l(x_i)$ is the estimated equated score for the l-th replicate.

Because the replicated samples are taken directly from the raw score data, this method is called empirical bootstrap. A variant of the empirical bootstrap utilizes a parametric model to sample from. Thus, in the parametric bootstrap (Efron and Tibshirani 1993) a parametric model is fitted to the data and is treated as if it describes the population appropriately so that the statistics of interest are simulated from the fitted model. Note that sampling with or without replacement is considered the same because the populations are assumed to have infinite size. To obtain parametric bootstrap standard errors, one can follow these steps:

1. Use log-linear modelling described in Sect. 2.3 to fit the empirical distributions of the test forms X and Y.
2. Use the fitted distributions from step 1 as population distributions for test forms X and Y and randomly select n_X scores for test X and n_Y scores for test Y. The obtained distributions are the parametric bootstrap sample distributions of scores on X and Y, respectively.
3. Estimate the equating transformation using the parametric bootstrap sample distributions of scores obtained in step 2.
4. Replicate step 2 and 3 L times.
5. Estimate the standard error (SE) at test score x_i over the samples using Eq. (3.27).

3.8.2 *Bias and RMSE*

Besides the standard error of equating, other measures such as bias and RMSE can be defined in the context of equating. For score x_i, let $\varphi(x_i)$ be the true equated score and $\hat{\varphi}_l(x_i)$ be the equated score based on sample l using any particular equating method. We define bias in a specific test score x_i as

$$\text{bias}(x_i) = \frac{1}{L}\sum_{l=1}^{L}[\hat{\varphi}_l(x_i) - \varphi(x_i)], \tag{3.28}$$

where L is the number of replicates (or samples). Combining the systematic (bias) and random standard errors (SE) yields the RMSE $= \sqrt{\text{bias}(x_i)^2 + \text{SE}(x_i)^2}$ in test score x_i, which can be estimated through

$$\text{RMSE}(x_i) = \sqrt{\frac{1}{L}\sum_{l=1}^{L}[\hat{\varphi}_l(x_i) - \varphi(x_i)]^2}. \tag{3.29}$$

Because "true" equated values do not exist, a criterion value $\varphi(x_i)$ is needed in order to estimate the bias. In the following section, examples using the **equate** package are shown to exemplify bootstrap standard errors, bias, and RMSE.

3.8.3 Examples Using equate

Estimates of SE, bias, and RMSE can be obtained using the `bootstrap()` function in the **equate** package, which implements both empirical and parametric bootstrap methods. The `bootstrap()` function is described later in this section. We start with an example showing how to obtain the bootstrap standard error of equating by adding the argument `boot=TRUE` in a call to the `equate()` function. In this case, bootstrap standard errors are based on sample sizes as specified in the arguments `xn` and `yn`. By default, the used sample sizes are matched with the observed sample sizes. The number of bootstrap samples is specified in the argument `reps`, which by default is set to 100 replications. To illustrate how to use bootstrap, we give an example using data from Kolen and Brennan (2014) and perform equipercentile equating using 100 bootstrap replications. The example shows a comparison between analytic standard errors and bootstrap standard errors. The values of the analytic standard errors are stored in the object `anse` and can be found in Kolen and Brennan (2014, p. 72). Reusing the objects `xACT` and `yACT` created earlier, the comparison is made by running the following code:

```
> req1 <- equate(xACT, yACT, type = "equipercentile",
+ boot = TRUE, reps = 100)

> anse <-c(1.9384,.8306, .5210, .8210, .2950, .1478, .2541,
+ .1582, .1969, .1761, .1731, .1952, .1800, .2311, .2431,
+ .2138, .2764, .2617, .3383, .2826, .2947, .3299, .3183,
+ .3865, .3555, .3013, .3683, .3532, .3069, .3422, .2896,
+ .3268, .3309, .3048, .3080, .3044, .3240, .2714, .3430,
+ .2018, .2787)
```

Figure 3.4 mimics Figure 7.1 in Kolen and Brennan (2014) and it was produced using the following code:

```
> plot(0:40,req1$bootstraps$se,type="l",lty=2,xlim=c(5,40),
+ ylim=c(0,0.5),xlab="Test X Score",ylab="Standard errors")
> lines(0:40,anse,lty=1)
> legend("topright",c("Bootstrap","Analytic"),lty=c(2,1))
```

It can be seen that bootstrap standard errors are very similar to analytic standard errors and are thus a valuable alternative when no formulas for standard errors are available. Note that because the bootstrap method involves random sampling, the appearance of the figure might slightly change if the analysis is redone.

To perform parametric bootstrap in **R**, one has to provide optional frequency distributions. Their role is to replace the sample distributions x and y when

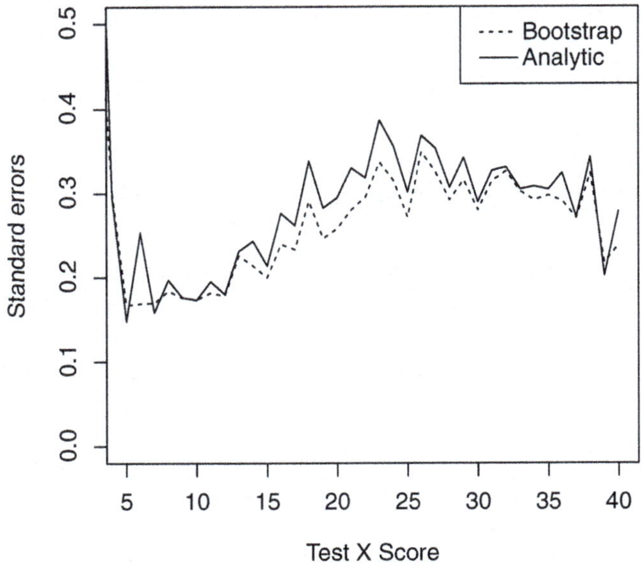

Fig. 3.4 Analytic and bootstrap standard errors

bootstrapping is performed (i.e. step 2 above). It is also possible to perform multiple equatings at each bootstrap replication with the `bootstrap()` function described in the next section. In **equate**, it is possible to obtain bias, standard errors, and RMSE for each equating function with the same bootstrap data. These ideas are illustrated in the next section.

3.8.4 *Additional Example: A Comparison of Traditional Equating Methods*

In this chapter a number of traditional equating methods have been described for different data collection designs. To illustrate how to compare and choose between equating methods, a comparison will be performed for the NEAT design using the ADM data set described in Chap. 2. This time, we will use parametric bootstrap where polynomial log-linear models are used to fit distributions as described in Sect. 2.3. Suppose that log-linear models chosen for X and Y with two univariate and one bivariate moments are adequate to fit the score data in the following way:

$$\log[p_{jl}] = \beta_0^X + \beta_1^X x_j + \beta_2^X x_j^2 + \beta_3^X a_l + \beta_4^X a_l^2 + \beta_5^{XA} x_j a_l, \tag{3.30}$$

and

$$\log[p_{kl}] = \beta_0^Y + \beta_1^Y y_k + \beta_2^Y y_k^2 + \beta_3^Y a_l + \beta_4^Y a_l^2 + \beta_5^{YA} y_k a_l. \tag{3.31}$$

Using the objects neat.x1 and neat.x2 created in Chap. 2, these log-linear models can be fitted using the following code:

```
> neat.xc <- presmoothing(neat.x1, "loglinear",
+ degrees = list(2, 1))
> neat.yc <- presmoothing(neat.y1, "loglinear",
+ degrees = list(2, 1))
```

To perform parametric bootstrap with the bootstrap() function, a few arguments need to be set. A typical call to the functions reads as

```
bootstrap(x, y, xn = sum(x), yn = sum(y), reps = 100,
crit, args, eqs = FALSE, ...)
```

where x and y are either score frequency distributions or previously created equating objects using equate(). The arguments xn and yn are the corresponding sample sizes which by default are set to the total observed number of test takers that have been administered test forms X and Y, respectively. The number of replications L can be specified using the (reps) argument, and its default value is 100. As mentioned before, to obtain a measure of bias (and thus a measure of RMSE as well), a criterion equating function $\varphi(x_i)$ is needed. The criterion equated values can be provided using the (crit) argument. To be able to compare different methods in terms of SE, bias and RMSE, one can perform multiple equatings in a single bootstrap study. In order for this to work, the arguments for each equating must be combined into a single object using the list() function, which is read into the args argument of the bootstrap() function.

For example, suppose that criterion equated values are considered to come as a result of an equipercentile frequency estimation equating method. If we use the presmoothed score data, the criterion equated values can be obtained as follows:

```
> critF <- equate(neat.xc, neat.yc, "equip",
+ "frequency estimation")$conc$yx
```

Now, suppose we want to compare the Tucker, Levine, chained linear, frequency estimation, and chained equipercentile equating methods. A simple way to do this is to create a list containing arguments for each equating method. An example is given below.

```
> neat.argms <- list(i = list(type = "i"),
+ lt = list(type="lin", method = "tucker"),
+ ll = list(type="lin", method = "levine"),
+ lc = list(type="lin", method = "chain"),
+ ef = list(type="equip", method = "freq",smooth="log"),
+ ec = list(type="equip", method = "chain",smooth="log"))
```

With these arguments, the bootstrap function can be used as follows

```
> bootNEAT <- bootstrap(x=neat.xc, y=neat.yc ,xn = 100,
+ yn = 100, reps = 100, crit = critF, args = neat.argms)
```

At each replication, an estimated equating is performed and saved. The bias in Eq. (3.28) and the RMSE in Eq. (3.29) can be obtained if a vector of criterion

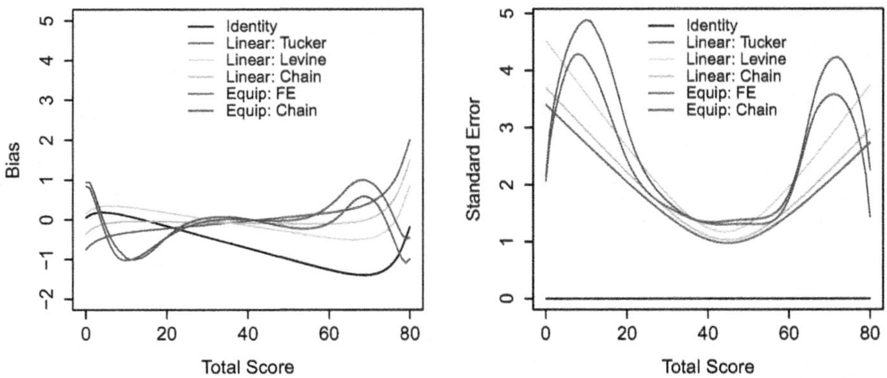

Fig. 3.5 Bias and standard errors for the five equating methods using the ADM data

equating scores is added via the argument `crit`, which in this example corresponds to frequency estimation equipercentile equating. To obtain a matrix of estimated equated values at each replication, the argument `eqs = TRUE` is used.

A more convenient way to analyze the results is by means of plots. We have chosen to show the obtained standard errors and bias in Fig. 3.5, although one can also obtain a plot of RMSE using the plot function as shown in the following code[3]:

```
> plot(bootNEAT, out = "se", addident = F,
+ col = c(1, rainbow(5)), legendplace = "top")
> plot(bootNEAT,out="bias",addident=F,legendplace="top",
+ col=c(1, rainbow(5)),morepars=list(ylim=c(-2, 5)))
> plot(bootNEAT,out="rmse",addident=F,legendplace="top",
+ col=c(1,rainbow(5)),morepars=list(ylim=c(0, 5)))
```

From Fig. 3.5, it can be seen that in terms of both bias (left panel) and standard errors (right panel), the linear Tucker equating seems to outperform the other equating methods.

These findings can be confirmed using the `summary()` function, which when applied to the `bootNEAT` object will show the bias, mean standard errors, RMSE, and absolute and weighted means from each equating. The weighted means are defined as the multiplication of the error estimate at each score point by the corresponding relative frequency in X. To obtain this information with the `summary` function, write

```
> round(summary(bootNEAT),2)
```

The obtained output[4] is the following table where the standard errors, bias and RMSE are given for the examined methods.

[3]Because of the random sampling involved in the bootstrap method, the obtained figures can be different.

[4]Because of the random sampling involved in the bootstrap method, the obtained output can be different.

```
    se w.se  bias a.bias w.bias wa.bias rmse w.rmse
i  0.00 0.00 -0.64   0.68  -0.01    0.01 0.68   0.01
lt 1.86 0.02  0.05   0.31   0.00    0.00 1.90   0.02
ll 2.44 0.02 -0.07   0.28   0.00    0.00 2.46   0.02
lc 2.01 0.02  0.00   0.15   0.00    0.00 2.02   0.02
ef 2.37 0.02 -0.02   0.38   0.00    0.00 2.41   0.02
ec 2.69 0.02 -0.15   0.36   0.00    0.00 2.73   0.02
```

It follows from the output that, indeed, the lowest value for standard errors is seen for the linear Tucker equating.

3.9 Summary

In this chapter traditional equating methods have been illustrated using the data described in Chap. 2. Several features of the **equate** package have been highlighted including the use of different equating methods for different data collection designs. In the last part of this chapter, it has been shown how to compare and assess different equating transformations using the bootstrap method for calculating standard errors and measures of bias and RMSE. A recent paper (Wiberg and González 2016) discusses more thoroughly how to assess an equating transformation, and these ideas will be discussed in detail in Chap. 7.

In the next chapter, kernel equating is introduced with examples using the **kequate** and **SNSequate** packages.

References

Albano, A. D. (2015). A general linear method for equating with small samples. *Journal of Educational Measurement, 52*(1), 55–69.

Albano, A. D. (2016). equate: An R package for observed-score linking and equating. *Journal of Statistical Software, 74*(8), 1–36.

Angoff, W. H. (1971). Scales, norms and equivalent scores. In R. L. Thorndike (Ed.), *Educational measurement* (2nd ed., pp. 508–600). Washington, DC: American Council on Education. Reprinted as Angoff W. H. (1984). *Scales, norms and equivalent scores*. Princeton, NJ: Educational Testing Service.

Babcock, B., Albano, A., & Raymond, M. (2012). Nominal weights mean equating a method for very small samples. *Educational and Psychological Measurement, 72*(4), 608–628.

Braun, H., & Holland, P. (1982). Observed-score test equating: A mathematical analysis of some ETS equating procedures. In P. Holland & D. Rubin (Eds.), *Test equating* (Vol. 1, pp. 9–49). New York: Academic Press.

Brennan, R. (2006). *Chained linear equating (casma technical note no. 3)*. Iowa City, IA: Center for Advanced Studies in Measurement and Assessment, University of Iowa.

Crocker, L., & Algina, J. (1986). *Introduction to classical and modern test theory*. New York: Holt, Rinehart, and Winston. ERIC.

Dorans, N. J. (1990). Equating methods and sampling designs. *Applied Measurement in Education, 3*(1), 3–17.

Efron, B. (1982). *The jackknife, the bootstrap and other resampling plans* (Vol. 38). Philadelphia: SIAM.

Efron, B., & Tibshirani, R. (1993). *An introduction to the bootstrap.* London: Chapman & Hall.
González, J. (2014). SNSequate: Standard and nonstandard statistical models and methods for test equating. *Journal of Statistical Software, 59*(7), 1–30.
Gulliksen, H. (1950). *Theory of mental tests.* New York: Wiley.
Hanson, B. A. (1991). A note on Levine's formula for equating unequally reliable tests using data from the common item nonequivalent groups design. *Journal of Educational and Behavioral Statistics, 16*(2), 93–100.
Hanson, B. A., Zeng, L., & Kolen, M. J. (1993). Standard errors of Levine linear equating. *Applied Psychological Measurement, 17*(3), 225–237.
Kolen, M., & Brennan, R. (2014). *Test equating, scaling, and linking: Methods and practices* (3rd ed.). New York: Springer.
Levine, R. (1955). *Equating the score scales of alternate forms administered to samples of different ability.* Research Bulletin 55–23. Princeton, NJ: Educational Testing Service.
Livingston, S., & Kim, S. (2009). The circle-arc method for equating in small samples. *Journal of Educational Measurement, 46*(3), 330–343.
Livingston, S. A., Dorans, N. J., & Wright, N. K. (1990). What combination of sampling and equating methods works best? *Applied Measurement in Education, 3*(1), 73–95.
Lord, F., & Novick, M. (1968). *Statistical theories of mental test scores.* Reading, MA: Addison-Wesley.
Petersen, N. S., Kolen, M. J., & Hoover, H. D. (1989). Scaling, norming, and equating. *Educational measurement, 3*, 221–262.
San Martín, E., & González, J. (2017). Analyzing the anchor test design in equating from an identification perspective (Manuscript submitted for publication).
Wiberg, M., & González, J. (2016). Statistical assessment of estimated transformations in observed-score equating. *Journal of Educational Measurement, 53*(1), 106–125.

Chapter 4
Kernel Equating

Abstract This chapter describes the kernel equating framework. The five steps that characterize kernel equating are illustrated using the `Math20EG`, `Math20SG`, `CBdata`, and `KB36` data sets that were introduced in Chap. 2 and which have been previously analyzed in the literature. We also illustrate the methods using the `ADM` admissions test data set. The **R** packages **kequate** (Andersson et al., J Stat Softw 55(6):1–25, 2013) and **SNSequate** (González, J Stat Softw 59(7), 1–30, 2014) are used throughout the chapter.

4.1 A Quick Overview of Kernel Equating

In the kernel equating approach (Holland and Thayer 1989; von Davier et al. 2004), continuous approximations of the discrete score distributions F_X and F_Y are obtained using kernel smoothing techniques (Silverman 1986). Such approximations are achieved by defining a *continuized* random variable that is a function of (i) the originally discrete score random variable, (ii) a continuous random variable characterizing the kernel, and (iii) a parameter controlling the degree of smoothness for the continuization (see Sect. 4.4). The conversion of scores is based on the estimated equating transformation

$$\hat{\varphi}(x; \mathbf{r}, \mathbf{s}) = F_{h_Y}^{-1}(F_{h_X}(x; \hat{\mathbf{r}}); \hat{\mathbf{s}}) = \hat{F}_{h_Y}^{-1}(\hat{F}_{h_X}(x)), \tag{4.1}$$

where h_X and h_Y are parameters that control the degree of smoothness in the continuization, and $\hat{\mathbf{r}}$ and $\hat{\mathbf{s}}$ are vectors of estimated score probabilities with coordinates defined as $r_j = \Pr(X = x_j)$ and $s_k = \Pr(Y = y_k)$ respectively, with x_j and y_k taking values in $\mathscr{X}$, and $\mathscr{Y}$, respectively. Both, $\hat{\mathbf{r}}$ and $\hat{\mathbf{s}}$ are obtained using the so-called *design functions* (DF), which take into account the chosen data collection design in the estimation (see Sect. 1.3.1). This process is performed after presmoothing the discrete (univariate and/or bivariate) observed-score frequency distributions by typically using log-linear models (see Sect. 2.3). The accuracy of the estimated $\hat{\varphi}(x)$ is assessed with different measures, especially the standard error of equating.

J. González, M. Wiberg, *Applying Test Equating Methods*,
Methodology of Educational Measurement and Assessment,
DOI 10.1007/978-3-319-51824-4_4

The kernel equating method has been summarized in the following five steps (see e.g., von Davier et al. 2004): (1) presmoothing, (2) estimation of scores probabilities, (3) continuization, (4) computation of the equating transformation, and (5) computation of accuracy measures. In the following sections, we provide details on the five steps in kernel equating together with examples of how to implement each of the steps in the **kequate** and **SNSequate** packages. Additional examples besides the ones shown in this chapter can be obtained from the book's webpage.

4.2 Step 1: Presmoothing

Presmoothing the score distributions as a first step when performing equating was described in Sect. 2.3 and the use of log-linear models for presmoothing and how to implement it with **equate** was described in Sect. 2.3.2. In this section, we describe how the presmoothing step in kernel equating is performed using both the **SNSequate** and **kequate** packages.

4.2.1 Presmoothing with SNSequate

Presmoothing in **SNSequate** is performed using the `loglin.smooth()` function. This function fits log-linear models to score data and provides estimates of the score probabilities (Step 2) as well as the **C** matrix[1] decomposition of their covariance matrix that is needed for the calculation of the standard error of equating (Step 5, see Sect. 4.6). The `loglin.smooth()` function is "design-specific", which means that its functionality will adapt according to the equating design specified in the argument `design` . A typical call to the function is as follows

```
loglin.smooth(scores, degree, design,...),
```

where the argument `scores` receives as input the score frequency distributions to be presmoothed, and the argument `degree` is used to specify the number of power moments to be fitted. In what follows, we give examples of univariate and bivariate presmoothing using the `Math20EG`, `Math20SG`, `CBdata`, and `KB36` data described in Chap. 2.

[1]Further details about **C** matrices are given in Appendix B.2.

4.2.1.1 Presmoothing Under the EG Design

We start with data collected under an EG design and fit a polynomial log-linear model for the X scores as described in Eq. (2.4). We fit a model with highest polynomial degree $T_r = 2$, which can be written as

$$\log(p_j) = \beta_0 + \beta_1^X(x_j) + \beta_2^X(x_j)^2. \tag{4.2}$$

The model in Eq. (4.2) can be fitted in **SNSequate** using the following code

```
> data(Math20EG)
> presEG <- loglin.smooth(scores = Math20EG[, 1],
+ degree = 2, design = "EG")
> presEG

Call:
loglin.smooth.default(scores = Math20EG[, 1],
+ degree = 2, design = "EG")

Estimated score probabilities:

   Score Est.Score.Prob.
1      0     0.002270957
2      1     0.004428770
3      2     0.008098301
4      3     0.013884856
5      4     0.022321610
6      5     0.033646997
7      6     0.047555830
8      7     0.063022827
9      8     0.078312061
10     9     0.091242272
11    10     0.099678200
12    11     0.102103574
13    12     0.098065977
14    13     0.088314586
15    14     0.074573268
16    15     0.059043294
17    16     0.043832338
18    17     0.030510924
19    18     0.019913736
20    19     0.012186718
21    20     0.006992903
```

The output shows the score scale in the `score` column, and the corresponding estimated score probabilities in the `Est.Score.Prob.` column. Note that these values coincide with what is reported in Table 7.4 in von Davier et al. (2004).

The object `presEG` contains the elements `sp.est`, `C`, and `psv`, which store the estimated score probabilities, $\hat{r}_j$, the $\mathbf{C}$ matrix, and the score scale $\mathscr{X}$, respectively. To exemplify the retrieving of these elements, suppose that we are interested in how the presmoothed score distribution compares to the original score distribution. Mimicking what is shown in Figure 7.1 in von Davier et al. (2004), Fig. 4.1 shows

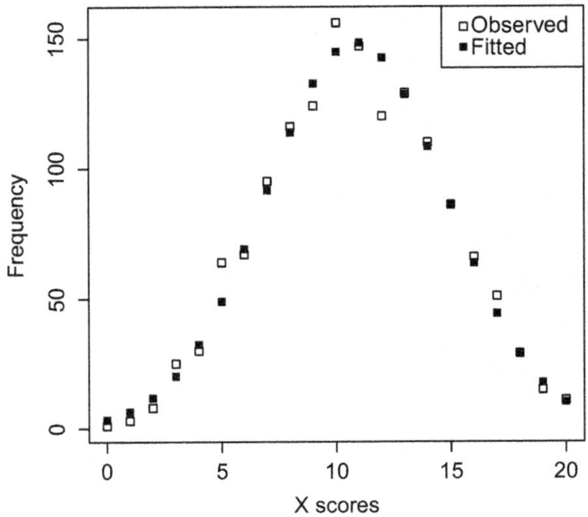

Fig. 4.1 The observed and the fitted distribution of X

a plot of the original and smoothed score distribution and was generated with the following code:

```
> data("Math20EG", package = "SNSequate")
> rj <- loglin.smooth(scores = Math20EG[, 1],
+ degree = 2, design = "EG")$sp.est
> sk <- loglin.smooth(scores = Math20EG[, 2],
+ degree = 3, design = "EG")$sp.est
> score <- 0:20
> plot(score, Math20EG[, 1],pch = 0,ylab="Frequency",
+ xlab = "X scores")
> points(score, rj * 1453, pch = 15)
> legend("topright",pch=c(0,15),c("Observed","Fitted"))
```

4.2.1.2 Presmoothing Under the SG Design

As was seen in Chap. 2, presmoothing under the SG design is possible by fitting a bivariate log-linear model for test scores X and Y. A convenient strategy is to start estimating the two univariate marginal score distributions for tests X and Y in the same way as for the EG design. When satisfactory univariate models are obtained, bivariate models should be estimated. Besides the univariate moments from each univariate log-linear model, the cross-moments between X and Y should be included up to the highest moments included in the univariate log-linear models. The best fitting model should be chosen according to the criteria discussed in Sect. 4.2.3. To presmooth the bivariate frequency distribution according to the SG design, we can, for example, consider the use of three power moments for each (marginal) X and Y score and one cross moment XY and fit the following log-linear model

$$\log(p_{jk}) = \beta_0 + \sum_{i=1}^{T_r=3} \beta_i^X (x_j)^i + \sum_{i=1}^{T_s=3} \beta_i^Y (y_k)^i + \sum_{i=1}^{L_r=1} \sum_{l=1}^{L_s=1} \beta_{il}^{XY} x_j^i y_k^l. \quad (4.3)$$

When `loglin.smooth()` is used to fit a bivariate score distribution under the SG design, the argument `degree` becomes a vector with entries equal to the number of power moments to be fitted to the marginal distributions and the number of cross moments to be fitted to the joint distributions.

```
> data(Math20SG)
> presSG<-loglin.smooth(scores = Math20SG,
+ degree = c(3, 3, 1, 1),design = "SG")
> presSG

Call:
loglin.smooth.default(scores = Math20SG,
+ degree = c(3, 3, 1, 1), design = "SG")

Estimated score probabilities:

   Score           r           s
1      0 0.001583093 0.001578752
2      1 0.003561066 0.003623786
3      2 0.007203015 0.007473589
4      3 0.013230277 0.013976282
5      4 0.022240753 0.023870355
6      5 0.034421089 0.037425573
7      6 0.049260072 0.054058035
8      7 0.065405878 0.072113498
9      8 0.080796899 0.089016839
10     9 0.093091894 0.101848078
11    10 0.100276704 0.108179978
12    11 0.101221187 0.106838498
13    12 0.095968537 0.098250225
14    13 0.085656941 0.084233492
15    14 0.072132287 0.067359828
16    15 0.057425354 0.050197807
17    16 0.043288323 0.034746895
18    17 0.030921249 0.022198047
19    18 0.020916943 0.012962520
20    19 0.013366505 0.006835540
21    20 0.008031934 0.003212383
```

In this case, `degree = c(3, 3, 1, 1)` means that $T_r = 3$, $T_s = 3$, $L_r = 1$, and $L_s = 1$, respectively. Note that although we have fitted a bivariate loglinear model, the output shows the marginal estimated score probabilities. This is due to the fact that `loglin.smooth()` internally applies the DF to obtain $\hat{r}_j$ and $\hat{s}_k$ according to the specified equating design (see Sects. 4.3 and B.1). Note also that the outputs coincide with the results shown in Table 8.5 in von Davier et al. (2004).

4.2.1.3 Presmoothing Under the CB Design

Presmoothing the data to be used under the CB design in **SNSequate** is possible by adding some additional arguments to the `loglin.smooth()` function. As we have seen in Sect. 1.3.1.3, the collected data under the CB design result in two independent bivariate vectors of scores. The argument `scores` is used for scores from the group taking test form X first and then later test form Y, whereas `scores2` is used for the group taking test form Y first and then test form X. We use the `CBdata` introduced in Chap. 2 to exemplify presmoothing under the CB design. `CBdata` is a list of two elements, `datx1y2` and `datx2y1`, containing the scores for the first and second group of test takers, respectively.

Suppose that for each of the bivariate score vectors, we want to fit the model

$$\log(p_{jk}) = \beta_0 + \sum_{i=1}^{T_r=2} \beta_i^X (x_j)^i + \sum_{i=1}^{T_s=2} \beta_i^Y (y_k)^i + \beta_{il}^{XY} x_j^i y_k^l. \tag{4.4}$$

The model in Eq. (4.4) can be fitted using the following code

```
> data(CBdata)
> presCB <- loglin.smooth(scores=CBdata$datx1y2,
+ degree=c(2,2,1,1,2,2,1,1), design="CB",
+ scores2=CBdata$datx2y1,K=76,J=77,wx=0.5,wy=0.5)$sp.est
```

The arguments `K` and `J` are used to indicate the number of possible test score values for test forms X and Y, respectively. The arguments `wx` and `wy` correspond to weights used within the CB design (see Sect. B.1) when the collected data are used as coming from two independent SG designs (see Sect. 1.3.1.3). The output (not shown) once again corresponds to the estimated score probabilities.

4.2.1.4 Presmoothing Under the NEAT Design

Kernel equating can be seen as an improved version of the traditional equipercentile equating in that percentile ranks and linear interpolation are replaced by more sophisticated smoothing techniques. As such, the assumptions made and the methods adopted for equipercentile equating methods under the NEAT design described in Sect. 3.4.2, can be translated to the kernel equating framework. As a matter of fact, both chained and frequency estimation kernel equating can be used under the NEAT design (see Sect. 4.5.1). In the kernel equating framework, frequency estimation is known as *poststratification* equating (PSE) (von Davier et al. 2004).

Under a NEAT design, the presmoothing step includes the fitting of two bivariate log-linear models. Let x_j and y_k be possible scores on test forms X and Y, respectively, and a_l be a possible score on the anchor test A. The score probabilities to be modeled using bivariate log-linear models are defined as

$$p_{jl} = \Pr(X = x_j, A = a_l), \tag{4.5}$$

and

$$q_{kl} = \Pr(Y = y_k, A = a_l), \tag{4.6}$$

where the p_{jl} probabilities are defined on population P and the q_{kl} probabilities are defined on population Q.

Suppose we want to fit the following log-linear model for X (and similarly for Y)

$$\log(p_{jl}) = \beta_0 + \sum_{i=1}^{T_r=2} \beta_i^X (x_j)^i + \sum_{i=1}^{T_s=2} \beta_i^A (a_l)^i + \beta_{il}^{XA} x_j^i a_l^l. \tag{4.7}$$

If the chained equating (CE) method is chosen, then we need to presmooth four different score distributions (see Eq. (3.24)). Accordingly, four vectors of score probabilities $\mathbf{r}_P$, $\mathbf{t}_P$, $\mathbf{s}_Q$, and $\mathbf{t}_Q$ have to be estimated (see Eq. (B.6)). The loglinear model in Eq. (4.7) can be fitted using the following code:

```
> presNEAT_CE<-loglin.smooth(scores=KBneat$x,
+ degreeXA=c(2,2,1,1),degreeYA=c(2,2,1,1),
+ design="NEAT_CE",scores2=KBneat$y,K=37,J=37,L=13)$sp.est
```

The arguments `degreeXA` and `degreeYA` function similarly to `degree` when `design="SG"` is selected (see Sect. 4.2.1.2). The values of `K`, `J`, and `L` indicate the total number of possible scores that can be obtained in test forms X, Y, and A, respectively. The displayed output corresponds to the estimated score probabilities.

On the other hand, if PSE is to be conducted, the following code can be used to fit the model in Eq. (4.7)

```
> presNEAT_PSE<-loglin.smooth(scores=KBneat$x,
+ degreeXA=c(2,2,1,1),degreeYA=c(2,2,1,1),
+ design="NEAT_PSE",scores2=KBneat$y,K=37,J=37,L=13,w=1)
```

The additional argument w is used here to specify the value of w_P in Eq. (3.4). The displayed output corresponds to the estimated score probabilities.

4.2.1.5 Modeling Complexities in the Data

The data sets used so far to illustrate the presmoothing step do not show any particular irregularities. Sometimes, however, the observed distributions are complex. Two common complexities in test data are discussed in von Davier et al. (2004, Chap. 10) and are briefly discussed here. The first is "teeth" or "gaps" in the observed score frequencies that occur at regular intervals due to the fact that scores are rounded to integers. The second is a "lump" or "spike" at 0 in the marginal distributions, and this occurs because negative values[2] are rounded to 0.

[2] Negative score values can arise for instance when in a multiple choice test, a fraction of the total score in the test is discounted for each wrong answer in the test.

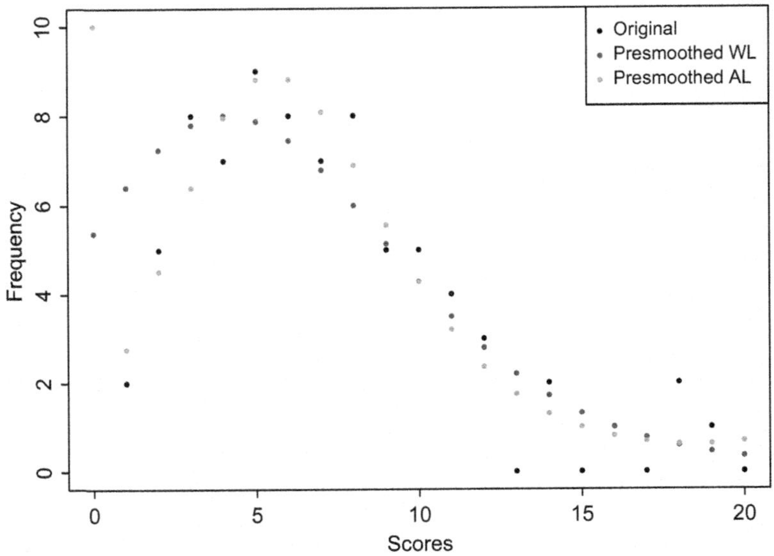

Fig. 4.2 Score frequencies distribution: original, presmoothed without accounting for the lump (WL), and accounting for the lump (AL)

One way to model lumps and teeth is to extend the elementary log-linear model by adding indicator variables for the particular 0 and teeth values. For a theoretical justification of this practice, see von Davier et al. (2004).

The `loglin.smooth()` function can handle these types of features in the data. To illustrate its use, we reproduce an example appearing in Moses and von Davier (2005). Figure 4.2 shows a score frequency distribution with a lump at $X = 0$ and compares this distribution with presmoothed versions that do not account for the lump (WL) and that do account for it (AL). The figure was produced using the following code

```
> score <- 0:20
> freq <- c(10, 2, 5, 8, 7, 9, 8, 7, 8, 5, 5, 4, 3, 0,
+ 2, 0, 1, 0, 2, 1, 0)
> ldata <- data.frame(score, freq)

> plot(ldata, pch=16, xlab="Scores",ylab="Frequency",
+ main="",col=1)
> m1 = loglin.smooth(scores=ldata$freq,kert="gauss",
+ degree=c(3),design="EG")
> m2 = loglin.smooth(scores=ldata$freq,kert="gauss",
+ degree=c(3),design="EG",lumpX=0)
> Ns = sum(ldata$freq)
> points(0:20,m1$sp.est*Ns, col=2, pch=16)
> points(0:20,m2$sp.est*Ns, col=3, pch=16)
> legend("topright",pch=c(16,16,16),col=c(1,2,3),
+ c("Original","Presmoothed WL","Presmoothed AL"))
```

4.2.2 *Presmoothing with kequate*

In Sect. 2.3.1, the observed score frequencies were assumed to follow a multinomial distribution so that maximum likelihood estimates of r_j and s_k can be calculated. A standard result of probability theory establishes that the conditional distribution of independent Poisson distributed random variables given the sum of these follows a multinomial distribution. Modeling Poisson data in **R** with the `glm()` function is straightforward. If we assume that the score frequencies follow a Poisson distribution, then dividing the obtained fitted values $\hat{n}_j$ by the sum of the frequencies allows estimates of r_j and s_k to be obtained directly.[3] This approach is adopted in the **kequate** package, which makes use of the `glm()` function to model the score frequencies. The created `glm()` object is read into **kequate** where further calculations are made internally. We thus first discuss briefly how the `glm()` function is used to estimate polynomial loglinear models. Using the resulting estimated score frequencies to obtain estimations of the score probabilities is discussed in Sect. 4.3.2. The `ADM` data set is used in the following sections, and additional examples using other data sets can be found in the book's webpage.

4.2.2.1 Presmoothing Under the EG Design

Consider the object `egADMk.x` created in Chap. 2 using the `kefreq()` function from **kequate**. Assume that the score data are well described by a polynomial log-linear model of the form

$$\log(p_j) = \beta_0 + \sum_{i=1}^{T_r=6} \beta_i^X (x_j)^i. \tag{4.8}$$

The following code can then be used to fit the model in Eq. (4.8) using the `glm()` function in **R**:

```
> egADMx <- glm(frequency~I(X)+ I(X^2)+
+ I(X^3)+ I(X^4)+ I(X^5)+I(X^6),
+ family = "poisson", data = egADMk.x, x = TRUE)
```

Because the ^ operator has a different meaning in an **R** formula, the `I()` is used to wrap polynomial terms. A summary of the fitted model appears as follows:

```
> summary(egADMx)

Call:
glm(formula = frequency ~ I(X) + I(X^2) + I(X^3) + I(X^4) +
 I(X^5) + I(X^6), family = "poisson", data = egADMk.x, x = TRUE)
```

[3] Note that this is only true for the EG design. For other designs, the vector of score probabilities has to be further transformed using the design functions in order to obtain $\hat{r}_j$ and $\hat{s}_k$.

```
Deviance Residuals:
     Min        1Q    Median        3Q       Max
-2.50354  -0.64481  -0.00515   0.47955   1.94570

Coefficients:
              Estimate Std. Error z value Pr(>|z|)
(Intercept) -1.657e+01  1.957e+00  -8.467  < 2e-16 ***
I(X)         2.981e+00  3.479e-01   8.570  < 2e-16 ***
I(X^2)      -1.659e-01  2.437e-02  -6.806 1.00e-11 ***
I(X^3)       4.894e-03  8.633e-04   5.669 1.44e-08 ***
I(X^4)      -8.046e-05  1.637e-05  -4.914 8.93e-07 ***
I(X^5)       6.952e-07  1.583e-07   4.390 1.13e-05 ***
I(X^6)      -2.480e-09  6.127e-10  -4.047 5.19e-05 ***
---
Signif. codes:  0 *** 0.001 ** 0.01 * 0.05 . 0.1  1

(Dispersion parameter for poisson family taken to be 1)

    Null deviance: 9439.808  on 80  degrees of freedom
Residual deviance:   69.392  on 74  degrees of freedom
AIC: 509.49

Number of Fisher Scoring iterations: 6
```

The output shows the estimated values for βs together with standard errors and associated p-values (the columns `Estimate`, `Std. Error`, and `Pr(>|z|)`, respectively). Fit statistics such as the AIC and the residual deviance are also reported. Although this output is the usual one for a `glm` object, we are more interested in the obtained fitted values because they will be used to estimate the score probabilities. Also, the **C** matrices (see Sect. B.2) are of interest for the calculation of the standard error of equating. As pointed out earlier, **kequate** will make use of the `egADMx` `glm` object and will internally calculate these elements. Nevertheless, how to obtain estimated score probabilities under the EG using the fitted frequency values from the glm output will be illustrated in Sect. 4.3.2. Also, in Sect. 4.5.2 we will show how the `glm` object `egADMx` can be read into **kequate** together with a similar object created for test Y in order to obtain a kernel equating transformation.

4.2.2.2 Presmoothing Under the SG Design

Presmoothing score data under the SG design can be done in a very similar way. For instance, using the `sgADMxy` object created in Chap. 2 with the `kefreq()` function for the `ADM` data set, the log-linear model in Eq. (4.3) with $T_r = T_s = L_r = L_s = 1$ can be fitted using

```
> SGadmXY <- glm(frequency~I(X)+I(A)+I(X):I(A),
+ family = "poisson",data = sgADMxy, x = TRUE)
```

Similarly as with the EG design, the glm object `SGadmXY` can then be read into **kequate** to perform a kernel equating as will be shown in Sect. 4.5.2.

4.2.2.3 Presmoothing Under the CB Design

To presmooth score data under the CB design we proceed in the same way as for the SG design, except that separate log-linear models are fitted for the two independent groups. For instance, using the `CBdata` we first create score distributions using `kefreq()`, then we fit the log-linear model given in Eq. (4.3) as follows:

```
> data("CBdata",package="SNSequate")
> CBadmx1y2 <- kefreq(CBdata$datx1y2[,1],0:75,
+ CBdata$datx1y2[,2],0:76)
> CBadmx2y1 <- kefreq(CBdata$datx2y1[,1],0:75,
+ CBdata$datx2y1[,2],0:76)
> CBadmXY1 <- glm(frequency~I(X)+I(X^2)+I(X^3)+I(A)
+ I(A^2)+I(A^3)+ I(X):I(A),family = "poisson",
+ data = CBadmx1y2, x = TRUE)
> CBadmXY2 <- glm(frequency~I(X)+I(X^2)+I(X^3)
+ I(A)+I(A^2)+I(A^3)+I(X):I(A),family = "poisson",
+ data = CBadmx2y1, x = TRUE)
```

Note that although there is no anchor test involved in the CB design, the `kefreq()` function assigns the label `A` by default when creating any bivariate score frequency distribution. This is why the `A` appears in the `glm` formula, and the same is true for the `SGadmXY` `glm` object created above.

4.2.2.4 Presmoothing Under the NEAT Design

Suppose we want to fit the log-linear model in Eq. (4.7). Using the admissions test data, the code below can be used to presmooth the score distributions using the objects `neatk.x` and `neatk.y` created in Chap. 2.

```
> NEATvX <- glm(frequency~I(X)+I(X^2)+I(A)+I(X):I(A),
+ family = "poisson", data = neatk.x, x = TRUE)
> NEATvY <- glm(frequency~I(X)+I(X^2)+I(A)+I(X):I(A),
+ family = "poisson", data = neatk.y, x = TRUE)
```

Note that regardless of whether CE or PSE is used, the presmoothing step is done in the same way. This is because the `glm()` function is only being used to fit the observed score frequencies. The `kequate()` function will then receive the `glm` object as an input and will automatically account for the different methods to estimate the score probabilities.

In Chap. 2, plots of the marginal frequencies of X and A in population P and of Y and A in population Q were introduced to graphically illustrate score data under the NEAT design. This type of plot was the default for the `plot()` function when the `freqtab()` function in **equate** was used to create bivariate frequency distributions (see, e.g., Fig. 2.4). When we are working on a kernel equating project using only the **kequate** or **SNSequate** package and the **equate** package has not been loaded, it is still possible to obtain the type of plots just described. As a matter of fact, the objects `verb.x` and `verb.xa` previously created in Chap. 2 were used to produce

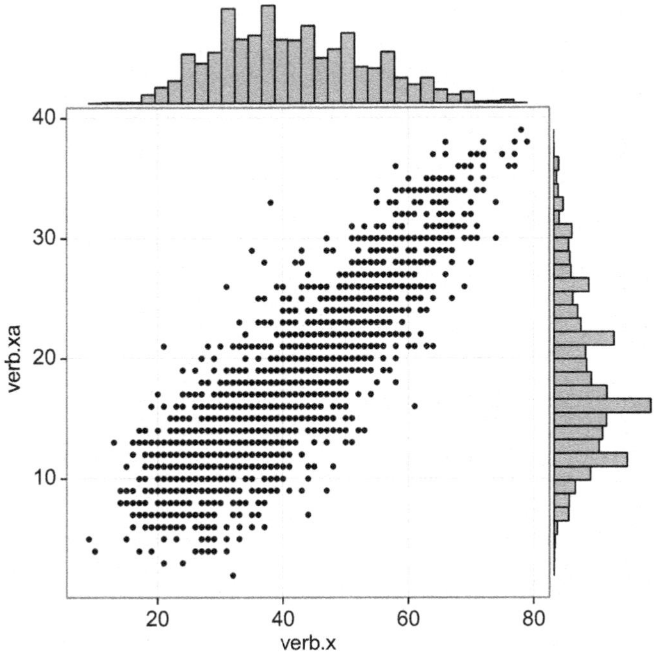

Fig. 4.3 Marginals of the bivariate frequency distributions for tests X and A in the ADM data under the NEAT design

Fig. 4.3 using the **R** packages **ggplot2** (Wickham 2009) and **ggExtra** (Attali 2016) and the following code:

```
> library(ggplot2)
> library(ggExtra)
> df1 <- data.frame(verb.x,verb.xa)
> (g1 <- ggplot(df1,aes(verb.x,verb.xa))
+ geom_point() + theme_bw())
> ggMarginal(g1, type = "histogram")
```

4.2.2.5 Modeling Complexities in the Data

To illustrate the modeling of complexities in the data when using the `glm()` function, we use the `ADM` data and the previously created object `neat.vx`. Note that because the `neat.vx` object was created using the `kefreq()` function, the data are already formatted in a proper way (i.e., ordered by the scores vectors). If the bivariate frequencies were not created using `kefreq()`, the `order` function in **R** can be used. To order a data frame first by the *A*-vector and then by the *X*-vector in the case of equal values of *A*, we can write

```
> neatXorder <- neatk.x[order(neatk.x$A, neatk.x$X),]
```

Assume that after examining the plots of the marginal frequencies, a lump is found at $X = 0$ and spikes are found at score values 5 and 10 for test form X. In order to incorporate this information in the log-linear model, we create indicator variables for each of these score values. We create two indicator variables, one that takes the value 1 if X is equal to 0 (`ix0`) and another that takes the value 1 if X has either the value 5 or 10. In practice we proceed as follows:

```
> neatk.x$ix0 <- numeric(length(neatk.x$X))
> neatk.x$ix1 <- numeric(length(neatk.x$X))
> neatk.x$ix0[neatk.x$X==0] <- 1
> neatk.x$ix1[neatk.x$X %in% c(5, 10)] <- 1
```

If the data contain more complexities in the X variable, we can create additional variables, and of course we can also construct new variables for particular values of the A and/or Y variables. A simple example that considers the fit of only the first two moments of the distribution of X, the first moment of the distribution of A, the cross moment XA, and the indicator variables defined previously is given in the code below.

```
> iNEAT <- glm(frequency~I(X)+I(X^2)+I(A)+I(X):I(A)
+ I(ix0)+I(ix1),data = neatk.x,
+ family = "poisson", x = TRUE)
```

4.2.2.6 Presmoothing Under the NEC Design

Presmoothing under a NEC design (Wiberg and Bränberg 2015) proceeds similarly as under the NEAT design. Recall from Chap. 1 that a NEC design typically uses other information from the test takers' covariates together with or instead of an anchor test. Because the covariates can be of any data type and not just integers, the specifications of the log-linear models can be somewhat more complicated. This means that the `kefreq()` function in **kequate** cannot always be used directly without manipulating the data first.

An example with the admissions test data with covariates is given here to illustrate how to presmooth data under a NEC design. Besides the test takers' scores, we have information about the two covariates sex (with values of 1 if male and 2 if female) and age category (with values of 1 if age is 20 years or younger, 2 if age is 21–24 years, 3 if age is 25–29 years, and 4 if age is 30 years or older). This means that there are $2 \times 4 = 8$ possible combinations of covariates. This also means that the frequency vector has length $81 \times 8 = 648$. The data are stored in the data frames `ADMnecX` and `ADMnecY`, each containing the test takers' scores and the covariates (sex and age). In this example we do not use the `kefreq()` function to create score distributions, and show how this process can be made in an alternative manner. The data are first sorted by age, then by sex, and finally by the test scores on test X. To obtain a data frame `NECxfreq` that contains the frequencies for all pairs of combinations of variables, we use the **R** functions `table()` and `as.data.frame()` as follows.

```
> load("ADMnecX.Rda")
> NECx <- data.frame(Xage=ADMnecX[,1], Xsex=ADMnecX[,2],
+ x=apply(ADMnecX[,43:122],1,sum))
> NECxfreq <- as.data.frame(table(factor(NECx$x,
+ levels = 0:80,ordered=TRUE),factor(NECx$Xsex,
+ levels = 1:2,ordered=TRUE),factor(NECx$Xage,
+ levels = 1:4,ordered=TRUE),dnn=c("necX","sex","age")))
```

Next, the test scores and covariates vectors have to be specified in a data frame using the `rep()` function. This ensures we have a data frame with correctly sorted vectors with all possible combinations.

```
> NECxdata <- data.frame(frequency = NECxfreq$Freq,
+ necX = rep(0:80,8), sex = rep(rep(1:2, each=81),2),
+ age = rep(1:4, each = 81*2))
```

When we have obtained a correctly sorted data frame, we can proceed to fit log-linear models in the same way as for the NEAT design. For instance, a polynomial log-linear model that fits three power moments of X, the main effects of the covariates, and the interaction of the X scores with each of the covariates can be obtained as follows.

```
> glmNECx <- glm(frequency~I(necX)+I(necX^2)+I(necX^3)+
+ factor(sex) + factor(age) + I(necX):I(sex)
+ I(necX):factor(age), data = NECxdata,
+ family = "poisson", x = TRUE)
```

This object is later used in kernel equating with a similarly constructed object for test Y.

No matter the chosen data collection design, several log-linear models should be fitted and compared in the presmoothing step. This is done in order to decide which model fits the data the best. Assessing model fit and selecting a good model for presmoothing is discussed in the next section.

4.2.3 Assessing Log-Linear Model Fit

There are several ways to assess and compare the fit of log-linear models. Moses and Holland (2009) recommend using the Akaike information criterion (AIC; Akaike 1974) because it is an effective criterion compared with other model selection strategies. Choosing a log-linear model using the AIC was described in Sect. 2.3.4.

Another strategy when assessing model fit is to examine residuals. A good choice would be to examine the Freeman-Tukey residuals (Bishop et al. 1975), which are approximately standard normal distributed if the observed frequencies are assumed to be Poisson distributed. If n_i is the i:th observed frequency and $\hat{n}_i$ is the i-th fitted frequency, then the Freeman-Tukey residuals are defined as

$$\mathrm{FT}_i = \sqrt{n_i} + \sqrt{n_i + 1} - \sqrt{4\hat{n}_i + 1}. \tag{4.9}$$

Unfortunately, analysis of the Freeman-Tukey residuals might be less useful in the bivariate case because there might be a large number of zero frequencies in the observed bivariate frequency distribution. When selecting a bivariate log-linear model, it is instead recommended to compare the models using AIC and likelihood ratio tests. A useful additional strategy to assess the dependency between the variables being fitted is to assess conditional parameters such as conditional means, variances, skewnesses, and kurtoses. The idea is that if the conditional parameters differ to a large extent between the observed and estimated distributions, one might need to try other models.

4.2.3.1 Assessing Log-Linear Model Fit in SNSequate

In **SNSequate** the `gof()` function contains various measures to assess the model's goodness of fit, and one of these measures is the Freeman-Tukey residuals. A typical function call reads as

```
gof(obs, fit, methods=c("FT"), p.out=TRUE)
```

where `obs` is a vector containing the observed values, and `fit` is a vector containing the fitted values. Supported options for `methods` are `FT` for Freeman-Tukey residuals (default), `Chisq` for Pearson's chi-squared test, and `KL` for the symmetrised Kullback-Leibler divergence (Johnson and Sinanovic 2000). The `p.out` argument is a logic that decides whether or not to display a QQ-plot of the Freeman-Tukey residuals. We use the `Math20EG` data to illustrate the `gof()` function. With the object `presEG` created in Sect. 4.2.1.1, the following code can be used to obtain the FT residuals, the symmetrised Kullback-Leibler divergence, and the chi-square statistic.

```
> assess.EG<-gof(Math20EG[,1], presEG$sp.est*1453,
+ method=c("FT", "KL", "Chisq"))
> assess.EG
Freeman-Tukey Residuals:
------------------------
 [1] -1.35391640 -1.43902355 -1.10463321  1.06028151
 [5] -0.38887200  2.04240224 -0.22359988  0.37997648
 [9]  0.22932204 -0.73407900  0.92992690 -0.09097184
[13] -1.94030685  0.08173209  0.18100165  0.04945792
[17]  0.31709466  0.99852452  0.05782508 -0.60222802
[20]  0.32760700

Symmetrised Kullback-Leibler divergence:
----------------------------------------
[1] 0.0132058

Pearson's Chi-squared Test:
---------------------------
X-squared = 420, df = 400, p-value = 0.236030325498846
---------------------------
```

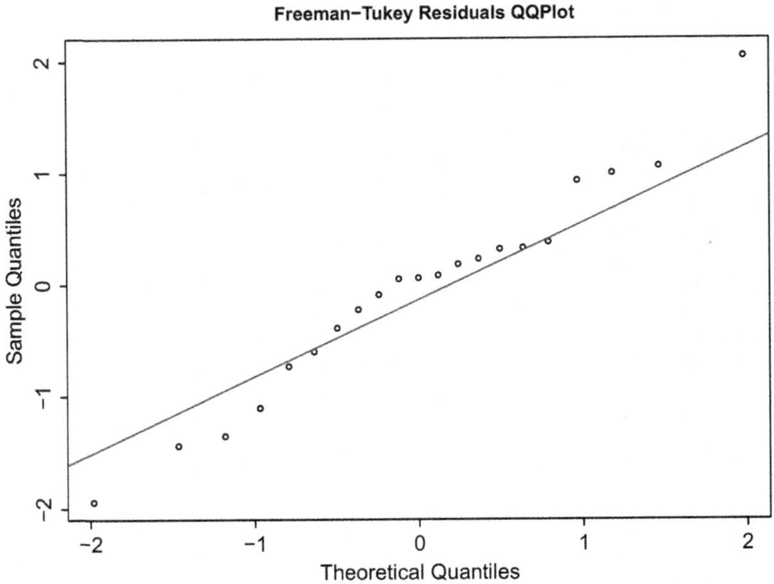

Fig. 4.4 QQ-plot for Freeman-Tukey residuals

By default, when `method="FT"` a QQ-plot is also displayed as output, and this is shown in Fig. 4.4. From the output, we can conclude that the model is appropriate for fitting the data.

4.2.3.2 Assessing Log-Linear Model Fit in kequate

The function `FTres()` in **kequate** is implemented to calculate the Freeman-Tukey residuals from an estimated log-linear model. The `FTres()` function receives as input two vectors containing the observed and estimated frequencies from the fitted log-linear model. The code below shows an example using the `Math20SG` data

```
> dataSG <- data.frame(X=rep(0:20, 21),
+ Y=sort(rep(0:20, 21)), frequency=as.numeric(t(Math20SG)))
> fitSG <- glm(frequency~I(X)+I(X^2)+I(X^3)+I(Y)+I(Y^2)+
+ I(Y^3)+I(X):I(Y), data = dataSG, family = "poisson",
+ x = TRUE)
> obsSG <- Math20EG
> estSG <- matrix(as.numeric(fitSG$fitted.values),
+ nrow= 21, byrow = T) / sum(Math20SG)

> SG.estX<-apply(estSG,1,sum)*1453
> SG.estY<-apply(estSG,2,sum)*1453
> SG.obsX<-apply(obsSG,1,sum)*1453
> SG.obsY<-apply(obsSG,2,sum)*1453
```

```
> FTx<-FTres(SG.obsX,SG.estX)
> FTy<-FTres(SG.obsY,SG.estY)
> FTsg<-round(cbind(0:20,SG.estX,FTx,SG.estY,FTy),2)
> FTsg
          SG.estX   FTx SG.estY   FTy
 [1,]  0     2.30 -0.78    2.29 -2.19
 [2,]  1     5.17 -0.93    5.27 -3.70
 [3,]  2    10.47 -0.72   10.86  0.40
 [4,]  3    19.22  1.27   20.31  0.83
 [5,]  4    32.32 -0.37   34.68  1.06
 [6,]  5    50.01  1.88   54.38  0.38
 [7,]  6    71.57 -0.52   78.55  1.48
 [8,]  7    95.03  0.02  104.78 -0.45
 [9,]  8   117.40 -0.11  129.34 -0.90
[10,]  9   135.26 -0.97  147.99 -0.39
[11,] 10   145.70  0.86  157.19 -0.64
[12,] 11   147.07  0.01  155.24 -0.16
[13,] 12   139.44 -1.68  142.76  0.29
[14,] 13   124.46  0.42  122.39  0.43
[15,] 14   104.81  0.52   97.87 -0.58
[16,] 15    83.44  0.30   72.94  0.38
[17,] 16    62.90  0.42   50.49  0.79
[18,] 17    44.93  0.91   32.25  1.01
[19,] 18    30.39 -0.21   18.83  0.09
[20,] 19    19.42 -1.00    9.93 -0.22
[21,] 20    11.67 -0.12    4.67 -3.44
```

Note that this results coincide with what is shown in Table 8.3 in von Davier et al. (2004).

The conditional moments for observed and estimated bivariate frequency distributions can be obtained in **kequate** through the function `cdist()`, which takes as arguments the matrices containing estimated (`est`) and observed (`obs`) frequencies in the population. By writing

```
> distSG <- cdist(est=estSG, obs=obsSG, 0:20, 0:20)
> plot(distSG)
```

we obtain the object `distSG`, which contains four data frames with the conditional parameters of each distribution. A plot of the conditional and fitted conditional mean and variances is shown in Fig. 4.5, where the triangles represent the observed values and the squares the fitted values. Note that this plot reproduces the results shown in Figures 8.4–8.6 in von Davier et al. (2004).[4] The chosen log-linear model is appropriate if the conditional parameters of the estimated distribution compared with those of the observed distribution do not deviate much from each other.

[4]Although variances instead of standard deviations are plotted in the right panel of Fig. 4.5.

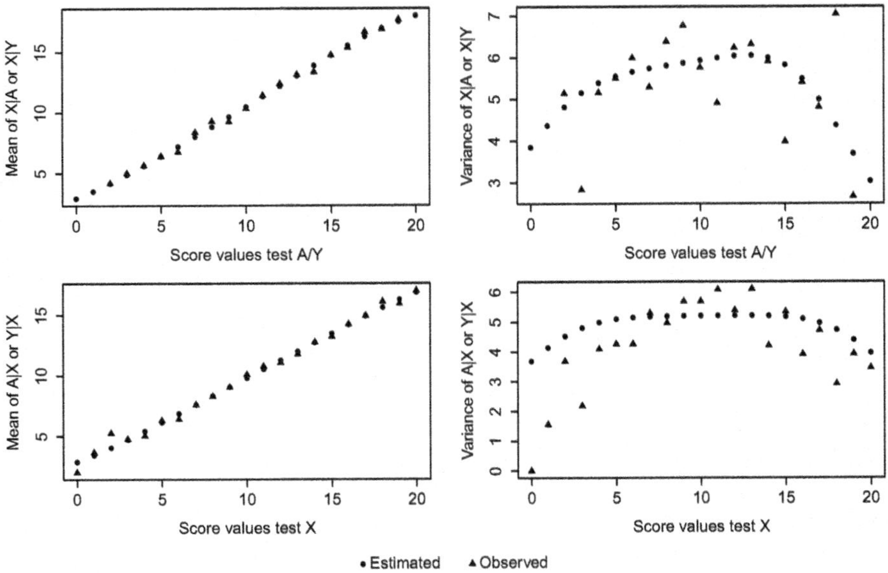

Fig. 4.5 The conditional parameters for each distribution using the `Math20SG`

4.3 Step 2: Estimation of Score Probabilities

Score probabilities are obtained from the presmoothed score distributions in step 1, and an important part in this step is the definition and the use of the design function (DF). The DF maps the estimated population score distributions into estimates of $\mathbf{r} = (r_1, r_2, \ldots, r_J)^t$ and $\mathbf{s} = (s_1, s_2, \ldots, s_K)^t$. The DF for the different data collection designs is given in Sect. B.1.

4.3.1 Estimation of Score Probabilities with SNSequate

Estimated score probabilities can be obtained as an output from the `loglin.smooth()` function. This function internally calculates both $\hat{\mathbf{r}}$ and $\hat{\mathbf{s}}$ according to the specified equating design. The code given in Sect. 4.2.1.1 for the EG design can be used to reproduce Table 7.4 in von Davier et al. (2004) if we write these additional lines

```
> Table7.4 <-cbind(score,rj,sk)
> Table7.4
```

that give the following output:

```
   score          rj          sk
1      0 0.002270957 0.001172655
2      1 0.004428770 0.002594327
3      2 0.008098301 0.005257340
4      3 0.013884856 0.009788287
5      4 0.022321610 0.016794173
6      5 0.033646997 0.026633849
7      6 0.047555830 0.039160187
8      7 0.063022827 0.053543058
9      8 0.078312061 0.068284368
10     9 0.091242272 0.081472531
11    10 0.099678200 0.091218940
12    11 0.102103574 0.096129091
13    12 0.098065977 0.095638497
14    13 0.088314586 0.090101256
15    14 0.074573268 0.080623425
16    15 0.059043294 0.068728368
17    16 0.043832338 0.055984437
18    17 0.030510924 0.043708630
19    18 0.019913736 0.032805583
20    19 0.012186718 0.023742208
21    20 0.006992903 0.016618788
```

4.3.2 *Estimation of Score Probabilities with kequate*

For an EG design, the DF is the identity function (see, Eq. (B.1)), thus the estimated score probabilities can be obtained through the estimated frequencies obtained from the `glm` output.[5] The following code shows an illustration using the `Math20EG` data

```
> egk.x <- kefreq(rep(0:20,Math20EG[,1]),0:20)
> egk.y <- kefreq(rep(0:20,Math20EG[,2]),0:20)
> EGx <- glm(frequency~I(X) + I(X^2),family = "poisson",
+ data = egk.x, x = TRUE)
> EGy <- glm(frequency~I(X) + I(X^2) + I(X^3),
+ family = "poisson", data = egk.y, x = TRUE)
> rj <- EGx$fitted.values/sum(EGx$fitted.values)
> sk <- EGy$fitted.values/sum(EGy$fitted.values)
```

[5]It might be possible to obtain estimated score probabilities from fitted frequencies for designs different than the EG, but additional steps might be necessary (i.e., the DF should be explicitly programmed and applied).

For the EG and all the other equating designs, **kequate** automatically estimates the score probabilities which can be easily retrieved using the function `getScores()`. If a kernel equating as described in Sect. 4.5.2 has been performed in **kequate**, the estimated score probabilities from the kernel equating object are retrieved using the `getScores()` function. The following code can be used to reproduce Table 7.4 in von Davier et al. (2004).

```
> MathEG <- kequate("EG",0:20,0:20,EGx,EGy)
> rj <- getScores(MathEG)$X$r
> sk <- getScores(MathEG)$Y$s
> Table7.4 <- cbind(0:20,rj,sk)
```

Instead of estimating values, **kequate** also allows the option to provide either vectors or matrices containing score probabilities. It is also possible to use observed proportions, which is useful when a kernel equating that omits the presmoothing step is performed.

4.4 Step 3: Continuization

As mentioned in the introduction of this chapter, continuizing a discrete score random variable involves the use of both a continuous random variable (characterizing the kernel to be used) and a *bandwidth* parameter (controlling the degree of smoothness in the continuization). In what follows, we describe continuization for the X score random variable, although the definitions for Y are analogous.

Let $X(h_X)$ be a continuized random variable defined as

$$X(h_X) = a_X(X + h_X V) + (1 - a_X)\mu_X,$$

where $\mu_X = \sum_j x_j r_j$, $\sigma_X^2 = \sum_j (x_j - \mu_X)^2 r_j$, $a_X = \sqrt{\sigma_X^2/(\sigma_X^2 + \sigma_V^2 h_X^2)}$, V is a continuous random variable characterizing the kernel, and h_X is a bandwidth parameter. Let $K(\cdot)$ be the chosen kernel, Theorem 4.2 in von Davier et al. (2004) shows that the kernel smoothing of F_X, is defined as

$$F_{h_X}(x) = \sum_j r_j K\left(\frac{x - a_X x_j - (1 - a_X)\mu_X}{a_X h_X}\right), \tag{4.10}$$

which is exactly the CDF of the continuized random variable $X(h_X)$. Although $K = \Phi$, the standard normal (or Gaussian) distribution function is a common choice of kernel, we will later see that other alternative kernels can also be used for equating.

4.4.1 Bandwidth Selection

The bandwidth h_X should be chosen so that the estimated density functions are smooth but the characteristics of the originally discrete score distributions are preserved. The bandwidth can be chosen in several ways. In von Davier et al. (2004) it is suggested to minimize the following penalty function

$$\text{PEN}\left(h_X\right) = \sum_j \left(\hat{r}_j - \hat{f}_{hX}\left(x_j\right)\right)^2 + k\sum_j \rho_j \tag{4.11}$$

where $\hat{f}_{hX}(x)$ is the estimated density function, k is a constant, and ρ_j is an indicator variable taking the value 1 if $\hat{f}'_{hX}(x_j - v) < 0$ and $\hat{f}'_{hX}(x_j + v) > 0$, and 0 otherwise. $\hat{f}'_{hX}(x_j)$ is the derivative of $\hat{f}_{hX}(x_j)$ and v is a parameter that indicates the width of the interval around the score x, typically chosen to be 0.25 (Lee and von Davier 2011; von Davier 2013). The first term in Eq. (4.11) preserves the characteristics of the distribution, although it could potentially undersmooth the data. The second term ensures the smoothness of the continuized distribution.

Alternative methods to choose the bandwidth in kernel equating have been proposed recently, including a cross-validation method (Liang and von Davier 2014), double smoothing (Häggström and Wiberg 2014), and a rule-based method (Andersson and von Davier 2014). Two of these new bandwidth selection methods are described in detail in Chap. 7.

4.4.2 Choosing the Kernel

Different choices besides a Gaussian kernel have been proposed in the literature. For example, Lee and von Davier (2011) explored the use of a uniform and a logistic kernel. If V is a uniform random variable and b is a positive real number, the uniform CDF is defined as

$$K(v) = \begin{cases} 0 & \text{for} \quad v < -b \\ (v+b)/2b & \text{for} \; -b \leq v < b. \\ 1 & \text{for} \quad v \geq b \end{cases} \tag{4.12}$$

The logistic CDF is defined as

$$K(v) = \frac{1}{1 + \exp(-v/s)}, \tag{4.13}$$

where s is a scale parameter and V has a mean of 0 and variance of $\sigma_V^2 = \pi^2 s^2/3$. These alternative kernels are implemented in both **kequate** and **SNSequate**.

Additionally, **SNSequate** also has implemented the Epanechnikov and adaptive kernels (Cid and von Davier 2015; González and von Davier 2017), which will be described in detail in Chap. 7.

In what follows, options for continuization (kernel and selection of bandwidth) are specified for **kequate** and **SNSequate** although examples are postponed for subsequent sections.

4.4.3 Continuization Choices in SNSequate

The function `bandwidth()` implements the minimization of the combined penalty function described in Eq. (4.11). A typical call reads as

```
bandwidth(scores, kert, degree, design, Kp = 1,
scores2, degreeXA, degreeYA, J, K, L, wx, wy, w, ...)
```

Many of the arguments function as explained in previous sections. The `Kp` argument correspond to the k constant in Eq. (4.11). Note that because the penalty function depends on both the estimated score probabilities[6] and the kernel density estimate, `bandwidth()` incorporates the `degree`, `design`, and `kert` arguments.

Choices of kernels are specified in the `ker.eq()` function using the argument `kert`. Current options for `kert` include the implementation of kernel equating using the Gaussian, logistic, uniform, Epanechnikov and adaptive kernels. In all of these cases, setting the bandwidth parameters can be done manually using the arguments `hx` and `hy`, or calculated automatically (default). In the last case, `ker.eq()` makes a call to the `bandwidth()` function.

4.4.4 Continuization Choices in kequate

The selection of bandwidth parameters can be done automatically or manually. The default is to use the penalty method in Eq. (4.11). Also, the arguments `hx` and `hy` in the `kequate()` function can be specified manually. In addition to the penalty function, two other bandwidth selection methods are currently implemented in **kequate**. The first is a rule-based selection method (Andersson and von Davier 2014) that only requires adding the argument `altopt=TRUE` in a call to `kequate()`. The second is an implementation of double smoothing (Hall et al. 1992) in the context of kernel equating (Häggström and Wiberg 2014) that only requires adding the argument `DS=TRUE` in a call to `kequate()`. These methods will be described in detail in Chap. 7.

[6]Estimated score probabilities are obtained by making a call to the `loglin.smooth()` function.

Currently, four different kernels are available in **kequate**: uniform, logistic, standard Gaussian, and Gaussian (default). These options are specified using the argument `kernel` in the `kequate()` function.

4.5 Step 4: Equating

Once the continuization step has been done, continuous approximations of the score distributions are available. Thus, the fourth step in kernel equating, which corresponds to the equating itself, can be done using the kernel equating transformation defined in Eq. (4.10), which we repeat here

$$\hat{\varphi}(x; \mathbf{r}, \mathbf{s}) = F_{h_Y}^{-1}(F_{h_X}(x; \hat{\mathbf{r}}); \hat{\mathbf{s}}) = \hat{F}_{h_Y}^{-1}(\hat{F}_{h_X}(x)).$$

4.5.1 Equating in SNSequate

To perform the actual equating in **SNSequate** the `ker.eq()` function is used. A typical call to `ker.eq()` has the following general structure

```
ker.eq(scores, kert, hx = NULL, hy = NULL, degree,
design, Kp = 1, scores2, degreeXA, degreeYA, J, K,
L, wx, wy, w, gapsX, gapsY, gapsA, lumpX, lumpY,
lumpA, alpha=NULL, h.adap=NULL)
```

where most of the arguments have already been explained in previous sections. The arguments `alpha` and `h.adap` will be explained in more details in Chap. 7 when adaptive kernel equating is introduced. Current options for `kert` are `gauss`, `logis`, `uniform`, `epan`, and `adap`. Possible options for the `design` argument are EG, SG, CB, `NEAT_CE`, and `NEAT_PSE`.

The examples that follow make use of the `Math20EG` and `CBdata` data. Other examples of kernel equating under the SG and NEAT designs can be found on the book's webpage.

We first illustrate kernel equating under the EG design by comparing the different results for: Gaussian, uniform and logistic kernels. In all cases, the log-linear models used in the presmoothing step are specified with $T_r = 2$ and $T_r = 3$ as given in Eq. (4.2)

```
> data("Math20EG")
> mod.logis <- ker.eq(scores=Math20EG, kert = "logis",
+ hx = NULL, hy = NULL, degree = c(2,3), design ="EG")
> mod.unif <- ker.eq(scores=Math20EG, kert="unif",
+ hx = NULL, hy = NULL, degree = c(2,3), design = "EG")
> mod.gauss <- ker.eq(scores=Math20EG, kert = "gauss",
+ hx = NULL, hy = NULL, degree=c(2,3), design="EG")
```

The output for the case of Gaussian kernel equating is as follows.

```
> mod.gauss

Call:
ker.eq.default(scores = Math20EG, kert = "gauss",
    hx = NULL, hy = NULL, degree = c(2, 3),
    design = "EG")

Equated values under the EG design:

   Score        eqYx       eqXy
1      0  0.3937085 -0.3215515
2      1  1.5812834  0.4964832
3      2  2.6403556  1.3862241
4      3  3.6443689  2.3557751
5      4  4.6316260  3.3604009
6      5  5.6177507  4.3748539
7      6  6.6099656  5.3870449
8      7  7.6120142  6.3912345
9      8  8.6259734  7.3847116
10     9  9.6530092  8.3661703
11    10 10.6934813  9.3353858
12    11 11.7471380 10.2925436
13    12 12.8126193 11.2385828
14    13 13.8868829 12.1751787
15    14 14.9641329 13.1051884
16    15 16.0338664 14.0334143
17    16 17.0781205 14.9679945
18    17 18.0676603 15.9235928
19    18 18.9607449 16.9287484
20    19 19.7182999 18.0476989
21    20 20.3929823 19.4152871
```

The displayed output shows the score scale (the `Score` column) and each of the equated values when X is equated to Y (the `eqYx` column) and when Y is equated to X (the `eqXy` column). A more complete output that includes summary statistics for scores, the selected bandwidth and kernel used in the continuization step, and the standard error of equating (see Sect. 4.6.1), is obtained by typing `summary(mod.gauss)`. These and other elements that are available from the object `mod.gauss` are listed in Table 4.1.

As an example of retrieving some of these elements, the following code can be used to reproduce Table 10.3 in Lee and von Davier (2011):

```
> XtoY <- cbind(LK.XtoY=mod.logis$eqYx,
+ UK.XtoY=mod.unif$eqYx,GK.XtoY=mod.gauss$eqYx)
> YtoX <- cbind(LK.YtoX=mod.logis$eqXy,
+ UK.YtoX=mod.unif$eqXy,GK.YtoX=mod.gauss$eqXy)
> Table10.3 <- round(cbind(XtoY,YtoX),3)
> Table10.3
```

Table 4.1 Elements stored in an object produced by `ker.eq()`

Element	Value
`kert`	Type of kernel (`gauss`, `logis`, `uniform`, `epan`, `adap`)
`design`	Equating design (`EG`, `SG`, `CB`, `NEAT_CE`, `NEAT_PSE`)
`eqYx`	Equated values when X is equated to Y
`eqXy`	Equated values when Y is equated to X
`h.x`	Selected bandwidth parameter for F_{h_X}
`h.y`	Selected bandwidth parameter for F_{h_Y}
`SEEYx`	Standard error of equating when X is equated to Y
`SEEXy`	Standard error of equating when Y is equated to X
`sevecYx`	SE-vector when X is equated to Y
`sevecXy`	SE-vector when Y is equated to X
`score`	Score scale
`rj`	Estimated score probabilities for X
`sk`	Estimated score probabilities for Y
`nx`	Number of test takers for X
`ny`	Number of test takers for Y
`meanx`	Mean of X
`meany`	Mean of Y
`sdx`	Standard deviation of X
`sdy`	Standard deviation of Y
`kurtx`	Kurtosis of X
`kurty`	Kurtosis of Y
`skewx`	Skewness of X
`skewy`	Skewness of Y
`de_f`	Density values $\hat{f}_{h_X}$
`de_g`	Density values $\hat{f}_{h_Y}$

```
        LK.XtoY UK.XtoY GK.XtoY LK.YtoX UK.YtoX GK.YtoX
 [1,]    0.447   0.439   0.394  -0.412  -0.227  -0.322
 [2,]    1.573   1.639   1.581   0.486   0.556   0.496
 [3,]    2.629   2.678   2.640   1.396   1.429   1.386
 [4,]    3.635   3.676   3.644   2.365   2.389   2.356
 [5,]    4.625   4.660   4.632   3.367   3.392   3.360
 [6,]    5.614   5.643   5.618   4.379   4.405   4.375
 [7,]    6.608   6.631   6.610   5.389   5.415   5.387
 [8,]    7.612   7.628   7.612   6.392   6.415   6.391
 [9,]    8.627   8.636   8.626   7.384   7.403   7.385
[10,]    9.655   9.658   9.653   8.365   8.379   8.366
[11,]   10.696  10.694  10.693   9.333   9.342   9.335
[12,]   11.750  11.745  11.747  10.290  10.295  10.293
[13,]   12.815  12.810  12.813  11.236  11.239  11.239
[14,]   13.888  13.885  13.887  12.174  12.175  12.175
[15,]   14.963  14.964  14.964  13.105  13.105  13.105
[16,]   16.031  16.035  16.034  14.034  14.033  14.033
[17,]   17.072  17.080  17.078  14.971  14.967  14.968
[18,]   18.058  18.073  18.068  15.930  15.920  15.924
```

```
[19,]  18.952  18.970  18.961  16.939  16.922  16.929
[20,]  19.734  19.707  19.718  18.058  18.036  18.048
[21,]  20.461  20.278  20.393  19.369  19.399  19.415
```

To perform kernel equating under other designs, we need to modify the argument `design` accordingly. Depending on the specified design, other arguments might be included in the `ker.eq()` function as well. To illustrate kernel equating under the CB design, we use the `CBdata`. Suppose that we want to estimate the kernel equating function when the same loglinear model as the one described in Sect. 4.2.1.3 is used in the presmoothing step and when the bandwidth parameters are selected automatically according to the penalty function method. The following code can be used:

```
> CBsns <- ker.eq(scores= CBdata$datx1y2, kert="gauss",
+ hx = NULL, hy = NULL, degree=c(2,2,1,1,2,2,1,1),
+ design="CB", Kp=0, scores2 = CBdata$datx2y1,
> J = 76,K = 77,wx=0.5,wy=0.5)
```

Using part of the output stored in the `CBsns` object, we can, for instance, plot the equating transformations for the cases when X is equated to Y and vice-versa, as was shown in Figure 9.10 in von Davier et al. (2004). Figure 4.6 shows such a plot and was produced using the following code:

```
> plot(0:75,CBsns$eqYx,pch=1,ylim=c(-6,78),
+ ylab="Equated Value",xlab="Score")
> points(0:76,CBsns$eqXy,pch=2)
> legend("topleft",pch=c(1,2),c("X to Y","Y to X"))
```

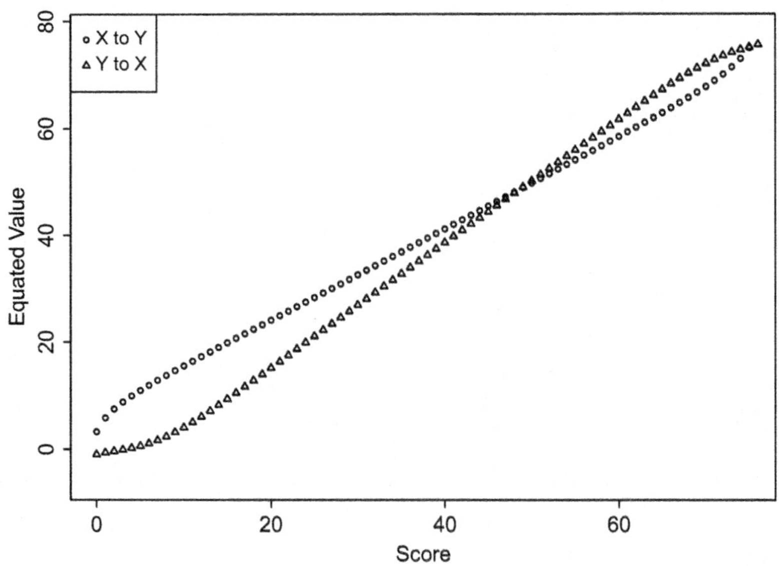

Fig. 4.6 Equating functions when X is equated to Y and vice-versa

4.5.2 *Equating in kequate*

The `kequate()` function is used to perform the actual kernel equating in **kequate**. A general function call is as follows:

```
kequate(design, ...)
```

where `design` can take the values `EG`, `SG`, `CB`, `EG`, `NEAT_CE`, `NEAT_PSE`, or `NEC`. Depending on the specified equating design, other arguments might need to be supplied. A list of the arguments that can be used in `kequate()` according to the selected design is shown in Table 4.2.

Table 4.2 Arguments supplied to `kequate()`

Argument(s)	Designs	Description
`x, y`	ALL	Score value vectors for test X and test Y.
`a`	CE	Score value vector for the anchor test A.
`r, s`	EG	Score probability vectors for tests X and Y. Alternatively objects of class `glm`
`P`	SG, CE, PSE, NEC	Matrix of bivariate score probabilities for tests X and Y (SG), tests X and A (CE, PSE), or test X and covariates (NEC) on population P. Alternatively an object of class `glm`
`Q`	CE, PSE, NEC	Matrix of bivariate score probabilities for tests Y and A (CE, PSE) or test Y and covariates (NEC) on population Q. Alternatively an object of class `glm`
`P12, P21`	CB	Matrices of bivariate score probabilities for tests X and Y. Alternatively objects of class `glm`
`DMP, DMQ`	CE, PSE, NEC	Design matrices for the specified bivariate log-linear models on populations P and Q, respectively (or groups taking test X and Y, respectively, in an EG design). Not needed if `P` and `Q` are of class `glm`
`DM`	SG	Design matrix for the specified bivariate log-linear model. Not needed if `P` is of class `glm`
`DM12, DM21`	CB	Design matrices for the specified bivariate log-linear models. Not needed if `P12` and `P21` is of class `glm`
`N`	ALL	The sample size for population P (or the group taking test X in the EG design). Not needed if `r`, `P`, or `P12` is of class `glm`
`M`	EG, CB, CE, PSE, NEC	The sample size for population Q (or the group taking test Y in the EG design). Not needed if `s`, `Q`, or `P21` is of class `glm`
`w`	PSE	Optional argument to specify the weight given to population P. Default is 0.5.
`hx, hy, hxlin, hylin`	EG, SG, CB, PSE, NEC	Optional arguments to specify the continuization parameters manually.

(continued)

Table 4.2 (continued)

Argument(s)	Designs	Description
hxP, hyQ, haP, haQ, hxPlin, hyQlin, haPlin, haQlin	CE	Optional arguments to specify the continuization parameters manually.
wcb	CB	The weighting of the two test groups in a counterbalanced design. Default is 0.5.
KPEN	ALL	Optional argument to specify the constant used in deciding the optimal continuization parameter. Default is 0.
wpen	ALL	An argument denoting at which point the derivatives in the second part of the penalty function should be evaluated. Default is 0.25.
linear	ALL	Logical denoting if only a linear equating is to be performed. Default is FALSE.
irtx, irty	ALL	Optional arguments to provide matrices of the probabilities of correctly answering the items on the parallel tests X and Y, as estimated in an IRT model.
smoothed	ALL	A logical argument denoting if the data provided are pre-smoothed or not. Default is TRUE.
kernel	ALL	A character vector denoting which kernel to use, with options "gaussian", "logistic", "stdgaussian", and "uniform". Default is "gaussian".
slog	ALL	The parameter used in the logistic kernel. Default is 1.
bunif	ALL	The parameter used in the uniform kernel. Default is 0.5.
altopt	ALL	Logical that sets the bandwidth parameter equal to a variant of Silverman's rule of thumb. Default is FALSE.
DS	EG, SG, CE, PSE, NEC	Logical that sets the bandwidth parameter equal to double smoothing. Default is FALSE.

The argument `linear` is a logical operator that if set to `TRUE` returns a linear equating version of the kernel transformation (see Theorem 4.5 in von Davier et al. 2004). The arguments `irtx` and `irty` are used to perform IRT kernel equating as described in Chap. 7. In all designs, different design matrices (e.g., `DM`, `DP`, `DMP`, `DMQ`, etc.) are needed if a `glm` object is not provided.

To illustrate kernel equating in `kequate()`, we use the `ADM` data and perform kernel equating under the EG, SG, NEAT, and NEC designs. The previously created objects `egADMx`, `egADMy`, `SGadmXY`, `NEATvX`, and `NEATvY` are used for the EG, SG and NEAT designs, respectively. For the NEC design, the previously created object `glmNECx` and a glm object `glmNECy`, similarly created using the same log-linear model, are used as input to `kequate()`. The uniform and logistic kernels are used for the EG and SG designs whereas the default Gaussian kernel is used for the other designs. The following code can be used to perform equating under the EG design.

```
> EGadm <- kequate("EG",0:80,0:80,egADMx,egADMy,
+ kernel = "uniform")
> summary(EGadm)
 Design: EG equipercentile

 Kernel: uniform

 Sample Sizes:
   Test X: 10000
   Test Y: 10000

 Score Ranges:
   Test X:
       Min = 0 Max = 80
   Test Y:
       Min = 0 Max = 80

 Bandwidths Used:
         hx        hy    hxlin    hylin
1 0.6478069 0.6365894 12718.08 12249.77

 Equating Function and Standard Errors:
   Score         eqYx      SEEYx
1      0 -0.29864646 0.04263775
2      1 -0.02288653 0.74487187
3      2  1.04369932 0.93949748
4      3  2.17929405 1.07861366
5      4  3.68737662 0.23059999
--    -------------------------
76    75 75.87150696 0.55516719
77    76 76.94807037 0.55545481
78    77 77.98642461 0.52804639
79    78 78.95953597 0.46192304
80    79 79.81899927 0.34549307
81    80 80.19076844 0.09109381

 Comparing the Moments:
        PREYx
1  -0.04160941
2  -0.08740705
3  -0.12721538
4  -0.15318140
5  -0.16490377
6  -0.16741158
7  -0.16719473
8  -0.16948332
9  -0.17729479
10 -0.19160971
```

The first part of the displayed output shows details on the performed equating such as the assumed equating design, the type of kernel used, and the bandwidth values used in the continuization step. The second part gives the score scale (the `Score` column), the equated values (the `eqYx` column), and the standard error of equating (the SEEYx column).[7] In the last part of the output, the percent relative error (the PREYx column) described in Sect. 4.6 is shown for the first ten moments. The output for the other designs that are examined below are similar and thus are omitted. The following can be used to perform equating under the SG, NEAT PSE, NEAT CE, and NEC designs, respectively.

```
> SGadm <- kequate("SG",0:80,0:80,SGadmXY, "logistic")
> neatPSEadm <- kequate("NEAT_PSE",0:80,0:80,
+ NEATvX, NEATvY)
> neatCEadm <-kequate("NEAT_CE",0:80,0:80,0:40,
+ NEATvX, NEATvY)
> NECadm <- kequate("NEC",0:80,0:80,glmNECx,glmNECy)
```

Among other useful elements, the created equating objects contain the equated values, which can be retrieved using the function getEq(). For instance, if we want to produce a table showing a comparison of the equated values obtained for the different equating designs we can use the following code:

```
> compare <- data.frame(EG=getEq(EGadm), SG=getEq(SGadm),
+ NEAT_PSE=getEq(neatPSEadm), NEAT_CE=getEq(neatCEadm),
+ NEC=getEq(NECadm))
> head(compare)
```

```
           EG        SG    NEAT_PSE     NEAT_CE         NEC
1 -0.29864646 0.1789797 -0.06991286 -0.07056532 -0.07139571
2 -0.02288653 1.4708836  0.86575669  0.86454785  0.87119888
3  1.04369932 2.7782494  1.81220941  1.81055003  1.83171535
4  2.17929405 4.0714894  2.77776014  2.77576969  2.81351072
5  3.68737662 5.3572239  3.75883075  3.75658998  3.80989961
6  4.72380634 6.6373788  4.75185071  4.74940945  4.81605625
```

Additional examples using other data sets and different options with **kequate** can be found on the book's webpage.

4.6 Step 5: Computation of Accuracy Measures

The last step in kernel equating is to assess the equating transformation by calculating accuracy measures such as the standard error of equating (SEE), the percent relative error (PRE), and the standard error of equating differences (SEED). For a detailed discussion on how an equating transformation should be assessed, refer to Wiberg and González (2016).

[7] Values from lines 6 to 75 have been omitted in order to save space.

4.6.1 Calculating the Standard Error of Equating

The SEE measures the uncertainty in the estimated equating transformation. The SEE for equating X to Y is denoted in general as

$$\mathrm{SEE}_Y(x) = \hat{\sigma}_Y(x) = \sqrt{\mathrm{Var}(\hat{\varphi}(x))} = \sqrt{\mathrm{Var}(\varphi(x; \hat{\mathbf{r}}, \hat{\mathbf{s}}))}. \tag{4.14}$$

As described in von Davier et al. (2004), the δ-method (Bishop et al. 1975; Lehmann 1999; Rao 1973; Kendall and Stuart 1997), which divides the problem into three parts, can be used to compute an estimate of the SEE. The first part consists of the Jacobian matrix $\mathbf{J}_\varphi$ of the equating transformation. The second part consists of the Jacobian matrix of the DF, $\mathbf{J}_{DF}$, which depends on the chosen data collection design. The final part relates to the asymptotic covariance matrix of the pre-smoothed frequencies obtained by the particular data collection design. After further calculations, which can be found in Sect. B.3, it can be shown that

$$\mathrm{SEE}_Y(x) = \left\| \mathbf{J}_\varphi \mathbf{J}_{DF} \mathbf{C} \right\|. \tag{4.15}$$

4.6.2 Standard Error of Equating Difference

To compare different kernel equating transformations, the SEED can be used. Let $\hat{\varphi}_1$ and $\hat{\varphi}_2$ represents two kernel equating transformations, then the SEED is defined as the Euclidean norm of the difference between two vectors of SEEs as follows[8]:

$$\mathrm{SEED}_Y(x) = \sqrt{Var(\hat{\varphi}_1(x) - \hat{\varphi}_2(x))} = \left\| \mathbf{J}_{\varphi_1} \mathbf{J}_{DF} \mathbf{C} - \mathbf{J}_{\varphi_2} \mathbf{J}_{DF} \mathbf{C} \right\|. \tag{4.16}$$

4.6.3 Percent Relative Error

Another way of comparing different equating transformations is through the moments of the score distributions. The PRE measures the difference between the moments of the distribution for equated values with those of the score distribution to which scores are being equated. The idea behind PRE is to get a measure of how well the equated scores match the observed distribution by comparing a number of moments. High values of PRE indicate a less effective score equating (Jiang et al. 2012). The PRE in the pth moment, PRE (p), is in general defined by

[8]The quantity inside the norm sign in Eq. (4.15) is called an SEE vector.

$$\mathrm{PRE}(p) = 100\frac{\mu_p(\varphi(X)) - \mu_p(Y)}{\mu_p(Y)}, \tag{4.17}$$

where $\mu_p(\varphi(X)) = \sum_j (\varphi(x_j))^p r_j$ and $\mu_p(Y) = \sum_k (y_k)^p s_k$ (von Davier et al. 2004).

4.6.4 Obtaining SEE, SEED, and PRE in SNSequate

The SEE is obtained as an output of the `ker.eq()` function. Both the SEE resulting from an equating of X to Y (`SEEYx`) and that resulting from an equating of Y to X (`SEEXy`) can be retrieved. For instance, using the previously created object `mod.gauss`, the following code shows how to retrieve the SEE for the equating of X to Y:

```
> seeYx<-mod.gauss$SEEYx
> seeYx
 [1] 0.22003271 0.28953356 0.28751030 0.26639755
 [5] 0.24104055 0.21695354 0.19666632 0.18124384
 [9] 0.17075038 0.16457168 0.16187109 0.16210071
[13] 0.16533595 0.17213358 0.18265321 0.19505015
[17] 0.20376009 0.19900477 0.16999780 0.11860165
[21] 0.07030529
```

Figure 4.7 shows a plot of the SEE produced using the following code:

```
> plot(score, mod.gauss$SEEYx, ylab = "SEEy(x)",
+ xlab = "X scores")
```

Both the SEE values and Fig. 4.7 show that values located in the middle of the score scale are more accurately equated. For this data set, this is an expected result because the frequency of extreme score values is lower. The figure mimics the one appearing as Figure 7.6 in von Davier et al. (2004).

The PRE and the SEED can be obtained in **SNSequate** with the functions `PREp()` and `SEED()`, respectively. Again, we use the `mod.gauss` object to exemplify the use of these functions. To obtain the PRE for the first 10 moments, we can write

```
> PREp(mod.gauss, 10)

Percent Relative Error:

  Moments      X_to_Y       Y_to_X
1       1 0.005861301 -0.006287818
2       2 0.012235716 -0.022772776
3       3 0.021915811 -0.054971354
4       4 0.039716757 -0.113883778
5       5 0.072771753 -0.213412807
6       6 0.127423113 -0.366390824
```

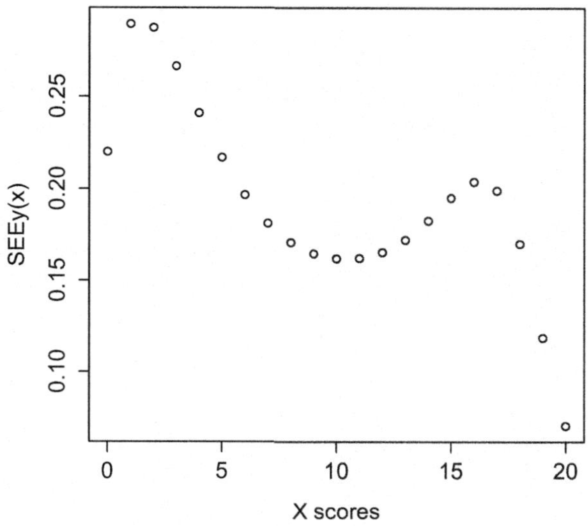

Fig. 4.7 SEE for the Math20EG data

```
7          7 0.208843818 -0.583677016
8          8 0.321024508 -0.873692528
9          9 0.466915023 -1.242324164
10        10 0.648615211 -1.693068922
```

From the obtained results, it can be seen by comparing the first five moments that the distribution of equated scores matches very well with the observed distribution of Y.

In Theorem 4.5 in von Davier et al. (2004) it is shown that a kernel equating is approximately linear when large values of bandwidth parameters are used. Suppose there is interest in comparing linear and Gaussian kernel equating. The following code can be used to obtain the corresponding SEED:

```
> mod.linear <- ker.eq(scores = Math20EG,kert = "gauss",
+ hx = 20, hy = 20, degree = c(2, 3), design = "EG")
> seedeg <- SEED(mod.gauss, mod.linear)$SEEDYx
> seedeg
 [1] 0.21120890 0.19421631 0.16553450 0.12978042
 [5] 0.09312270 0.05868800 0.02897404 0.01313716
 [9] 0.02531616 0.03841881 0.04544525 0.04544446
[13] 0.03869717 0.02799566 0.02516353 0.04100783
[17] 0.06276693 0.08238740 0.10573648 0.14917952
[21] 0.20243258
```

It can be seen that the standard error of equating differences range from 0.01 to 0.21. The differences between each of the equated values obtained using the Gaussian kernel equating function with those obtained using the approximated linear equating function can be assessed for significance by comparing them to their uncertainty (i.e., their SEED values). A graphical alternative, as shown in Figure 7.7 in von Davier et al. (2004), is to plot such differences of equated values along with ±2SEED values. Figure 4.8 shows an example generated using the following code:

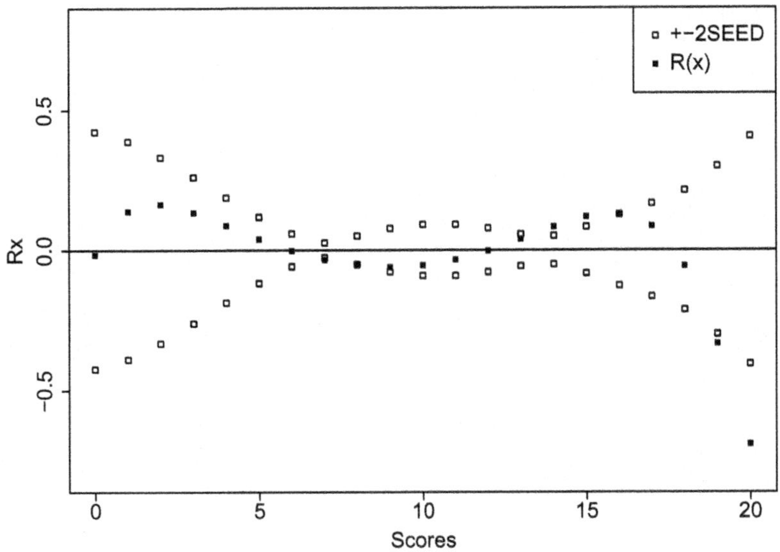

Fig. 4.8 Difference between linear and kernel equating functions ±2SEED

```
> Rx <- mod.gauss$eqYx - mod.linear$eqYx
> plot(0:20,Rx,ylim=c(-0.8,0.8),pch=15,xlab="Scores")
> abline(h = 0)
> points(0:20, 2 * seedeg, pch = 0)
> points(0:20, -2 * seedeg, pch = 0)
> legend("topright",pch=c(0, 15),c("+-2SEED", "R(x)"))
```

Other examples can be found on the book's webpage.

4.6.5 *Obtaining SEE, SEED, and PRE in kequate*

The SEE, SEED, and PRE can be obtained in **kequate** using the functions `getSee()`, `genseed()`, and `getPre()`, respectively. Also, the `getSeed()` function is used to obtain the SEED between a linear and a kernel equating function. These functions are applied to equating objects that have been created using the `kequate()` function. We use the `ADM` data with the previously created object `EGadm` to obtain the SEE and the PRE from a kernel equating under the EG design.

```
> AdmEGsee <- getSee(EGadm)
> AdmEGsee
 [1] 0.04263775 0.74487187 0.93949748 1.07861366
 [5] 0.23059999 0.27028746 0.30010687 0.31745325
 [9] 0.32169924 0.31397991 0.29666718 0.27275267
[13] 0.24530141 0.21705960 0.19022940 0.16637795
[17] 0.14643196 0.13072108 0.11906270 0.11090145
[21] 0.10549726 0.10211294 0.10014268 0.09916074
```

```
[25] 0.09891059 0.09926657 0.10549865 0.10525719
[29] 0.10586187 0.10728897 0.10950213 0.11244512
[33] 0.11519517 0.11830876 0.12194774 0.12598671
[37] 0.13031128 0.13482367 0.13944732 0.13884922
[41] 0.14293797 0.14710155 0.15133965 0.15565451
[45] 0.16004005 0.16447198 0.16890153 0.17325493
[49] 0.17743963 0.18135684 0.19559914 0.19917012
[53] 0.20230967 0.20506263 0.20752743 0.20984191
[57] 0.21215767 0.21460674 0.21726727 0.23937514
[61] 0.24356042 0.24780185 0.25187033 0.25554615
[65] 0.25868979 0.26131550 0.26365257 0.30936000
[69] 0.31652428 0.32596446 0.33833893 0.35366272
[73] 0.37093851 0.38794507 0.53690264 0.55516719
[77] 0.55545481 0.52804639 0.46192304 0.34549307
[81] 0.09109381
> AdmEGpre <- getPre(EGadm)
> AdmEGpre
        PREYx
1  -0.04160941
2  -0.08740705
3  -0.12721538
4  -0.15318140
5  -0.16490377
6  -0.16741158
7  -0.16719473
8  -0.16948332
9  -0.17729479
10 -0.19160971
```

The functions `getSee()`, `getSeed()`, and `genseed()` have implemented plot methods. When an object created by `getSee()` is passed to `plot()`, a plot showing the SEE for each score value will be displayed. In the case of `getSeed()` and `genseen()`, a plot showing the equated value differences together with the associated SEED will be displayed. An example is shown in Fig. 4.9 where the left-hand plot was obtained from

```
> plot(AdmEGsee, xlab ="X scores",ylab="SEE")
```

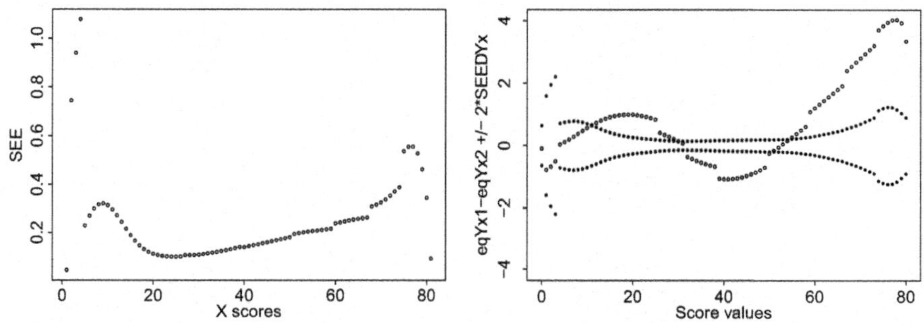

Fig. 4.9 SEE and SEED using the admissions data with an EG design

and the right-hand plot was obtained by writing

```
> AdmEGseed <- getSeed(EGadm)
> plot(AdmEGseed)
```

The `genseed()` function gives the SEED between two different **kequate** objects. For instance, we can compare the previously performed poststratification equating (`neatPSEadm`) to the chained equating (`neatCEadm`) using the admissions test data. The resulting output gives the difference between the equated values of the two equating functions (`eqYxD`) and the SEED (`SEEDYx`).[9]

```
> ADMseedNEAT <- genseed(neatPSEadm,neatCEadm)
> ADMseedNEAT
An object of class "genseed"
Slot "out":
          eqYxD      SEEDYx
1  0.0006524646 0.01363087
2  0.0012088376 0.02537221
3  0.0016593778 0.03827020
4  0.0019904405 0.04948587
5  0.0022407612 0.05914505
---
76 0.0264014324 0.08284398
77 0.0221884565 0.07003946
78 0.0176941815 0.05623997
79 0.0129488065 0.04151021
80 0.0083604288 0.02705509
81 0.0048851535 0.01571760
```

It is also possible to retrieve other useful elements stored in a `kequate()` object. A summary of the functions used to retrieve them is shown in Table 4.3. Further examples can be found in the book's webpage.

Table 4.3 Functions to retrieve information from the resulting `kequate()` objects

Function	Output
`getEquating()`	A data frame with the equated values, SEEs and other information about the equating.
`getPre()`	A data frame with the PRE for the equated distribution.
`getType()`	A character vector describing the type of equating conducted.
`getScores()`	A vector containing the score values for the equated tests.
`getH()`	A data frame containing the values of h used in the equating.
`getEq()`	A vector containing the equated values.
`getEqlin()`	A vector containing the equated values of the linear equating.
`getSeelin()`	A vector containing the SEEs for the equated values of the linear equating.
`getSeed()`	An object of class genseed containing the SEED between the kernel. equipercentile equating and the linear equating (if applicable).

[9]Values from lines 6–75 have been omitted to save space.

4.7 Different Features in kequate and SNSequate

For the implementation of the kernel equating framework, the **R** packages **kequate** and **SNSequate** share many similarities, but there are some important differences in what can currently be done in both of them.

The **kequate** package allows both observed frequencies and `glm` objects as inputs for score distributions. The latter option has been shown to be very useful and flexible to account for model complexities in the presmoothing step (see, Sect. 4.2.2.5), and to incorporate covariates under the NEC design (see Sect. 4.2.2.6). **kequate** also has specific functions like `cdist()` to assess the model fit, and others like the `get` functions shown in Table 4.3 to retrieve information about the performed kernel equating. **kequate** also has some other distinct features that are discussed in detail in Chaps. 6 and 7. Among these are the option to choose different bandwidth selection methods, and to perform IRT observed-score kernel equating and local kernel equating.

The **SNSequate** package on the other hand has implemented the Epanechnikov and adaptive kernels in addition to the Gaussian, uniform and logistic kernels. It also contains the function `gof()` that implements various measures to assess a model's goodness of fit which is especially useful in the presmoothing step.

4.8 Summary

In this chapter, the five steps of kernel equating were discussed and the **R** packages **SNSequate** and **kequate** were used to illustrate kernel equating under the SG, EG, CB, NEAT, and NEC designs. Both packages implement all of the procedures described in von Davier et al. (2004), although each package has specific additional features that have been discussed here. Further improvements in kernel equating that have also been implemented in either **SNSequate** or **kequate** will be discussed in Chap. 7.

References

Akaike, M. (1974). A new look at the statistical model identification. *IEEE Transactions on Automatic Control, 19*, 716–723.

Andersson, B., & von Davier, A. A. (2014). Improving the bandwidth selection in kernel equating. *Journal of Educational Measurement, 51*(3), 223–238.

Andersson, B., Bränberg, K., & Wiberg, M. (2013). Performing the kernel method of test equating with the package kequate. *Journal of Statistical Software, 55*(6), 1–25.

Attali, D. (2016). *ggExtra: Add marginal histograms to 'ggplot2', and more 'ggplot2' enhancements*. R package version 0.3.4.

Bishop, Y., Fienberg, S., & Holland, P. (1975). *Discrete multivariate analysis: Theory and practice*. Cambridge, MA: MIT Press.

Cid, J. A., & von Davier, A. A. (2015). Examining potential boundary bias effects in kernel smoothing on equating: An introduction for the adaptive and Epanechnikov kernels. *Applied Psychological Measurement, 39*(3), 208–222.

González, J. (2014). SNSequate: Standard and nonstandard statistical models and methods for test equating. *Journal of Statistical Software, 59*(7), 1–30.

González, J., & von Davier, A. A. (2017). An illustration of the Epanechnikov and adaptive continuization methods in kernel equating. In L. A. van der Ark, M. Wiberg, S. A. Culpeppe, J. A. Douglas, & W.-C. Wang (Eds.), *Quantitative psychology – 81st annual meeting of the psychometric society, Asheville, North Carolina, 2016*. New York: Springer.

Häggström, J., & Wiberg, M. (2014). Optimal bandwidth selection in observed-score kernel equating. *Journal of Educational Measurement, 51*(2), 201–211.

Hall, P., Marron, J., & Park, B. U. (1992). Smoothed cross-validation. *Probability Theory and Related Fields, 92*(1), 1–20.

Holland, P., & Thayer, D. (1989). *The kernel method of equating score distributions* (Technical report). Princeton, NJ: Educational Testing Service.

Jiang, Y., von Davier, A. A., & Chen, H. (2012). Evaluating equating results: Percent relative error for chained kernel equating. *Journal of Educational Measurement, 49(1)*, 39–58.

Johnson, D. H., & Sinanovic, S. (2000). *Symmetrizing the Kullback-Leibler distance* (Technical report). IEEE Transactions on Information Theory.

Kendall, M., & Stuart, A. (1997). *The advanced theory of statistics* (4th ed., Vol. 1). London: Griffin.

Lee, Y., & von Davier, A. (2011). Equating through alternative kernels. In A. von Davier (Ed.), *Statistical models for test equating, scaling, and linking* (Vol. 1, pp. 159–173). New York: Springer.

Lehmann, E. L. (1999). *Elements of large-sample theory*. New York: Springer.

Liang, T., & von Davier, A. A. (2014). Cross-validation: An alternative bandwidth-selection method in kernel equating. *Applied Psychological Measurement, 38*(4), 281–295.

Moses, T., & Holland, P. (2009). Selection strategies for univariate loglinear smoothing models and their effect on equating function accuracy. *Journal of Educational Measurement, 46*, 159–176.

Moses, T., & von Davier, A. (2005). Using PROC GENMOD for loglinear smoothing. Paper SA0605, 1–21.

Rao, C. R. (1973). *Linear statistical inference and applications*. New York: Wiley.

Silverman, B. (1986). *Density estimation for statistics and data analysis* (Vol. 3). London: Chapman and Hall.

von Davier, A. A. (2013). Observed-score equating: An overview. *Psychometrika, 78*(4), 605–623.

von Davier, A. A., Holland, P., & Thayer, D. (2004). *The kernel method of test equating*. New York: Springer.

Wiberg, M., & Bränberg, K. (2015). Kernel equating under the non-equivalent groups with covariates design. *Applied Psychological Measurement, 39*(5), 349–361.

Wiberg, M., & González, J. (2016). Statistical assessment of estimated transformations in observed-score equating. *Journal of Educational Measurement, 53*(1), 106–125.

Wickham, H. (2009). *ggplot2: Elegant graphics for data analysis*. New York: Springer.

Chapter 5
Item Response Theory Equating

Abstract In this chapter, different methods of Item Response Theory (IRT) linking and equating will be discussed and illustrated using the **SNSequate** (González, J Stat Softw 59(7):1–30, 2014) and **equateIRT** (Battauz, J Stat Softw 68(7):1–22, 2015) packages. Other useful packages include **ltm** (Rizopoulos, J Stat Softw 17(5):1–25, 2006) and **mirt** (Chalmers, J Stat Softw, 48(6):1–29, 2012), which allow the user to model response data using different IRT models. IRT objects obtained from the latter packages can also be read into **equateIRT** and **kequate** (Andersson et al., J Stat Softw, 55(6):1–25, 2013) to perform IRT equating and linking.

5.1 IRT Models

IRT models (Lord 1980; Hambleton and Swaminathan 1985; De Boeck and Wilson 2004; van der Linden 2016) are widely used nowadays for analyzing and scoring tests. Because many testing programs use IRT to assemble tests, the use of IRT equating is a natural choice for equating (Skaggs and Lissitz 1986; Cook and Eignor 1991; Lord 1980, Chap. 13). In what follows, we briefly describe the specification of IRT models.

Let X_{ij} be the random variable denoting the answer of individual i on item j in test form X (the notation and definitions that follow can easily be adapted for test form Y). Assuming $i = 1, \ldots, n_x$ test takers and $j = 1, \ldots, J_x$ items, the observed data can be accommodated in an $n_x \times J_x$ matrix where each row contains the response pattern of each test taker. Note that in the case where items are binary scored (i.e., 1 if the answer is correct and 0 otherwise), sum scores $X_i = \sum_{j=1}^{J_x} X_{ij}$ can be computed and used as the test taker's i score. An alternative way to produce test takers' scores is to use IRT models. IRT models for binary-scored items specify the respondent's probability of a correct answer on a test item based on both a person's parameter, θ_i, and a vector of item characteristics, ω_j (e.g., the difficulty level of the item, the

J. González, M. Wiberg, *Applying Test Equating Methods*,
Methodology of Educational Measurement and Assessment,
DOI 10.1007/978-3-319-51824-4_5

capacity of the item to discriminate among test takers, etc.). When it is assumed that $\theta_i \sim N(0, \sigma_\theta^2)$ the statistical model[1] becomes

$$(X_{ij} \mid \theta_i, \omega_j) \sim \text{Bernoulli}(\pi(\theta_i, \omega_j)), \tag{5.1}$$

where $\pi(\cdot)$ is known as the item characteristic curve (ICC). In particular, the three parameter logistic IRT model (3PL, Birnbaum 1968) employs $\pi(\theta_i, \omega_j) = c_j + (1 - c_j)\Psi(Da_j(\theta_i - b_j))$, where $\omega_j = (a_j, b_j, c_j)$, D is a scaling constant, and $\Psi(x) = \exp(x)/1 + \exp(x)$ is the standard logistic function. The a_j, b_j, c_j components in ω_j are the discrimination, difficulty, and guessing item parameters, respectively. Under this specification, we have

$$\pi_{ij} = \Pr(X_{ij} = 1 \mid \theta_i, \omega_j) = c_j + (1 - c_j)\frac{\exp[Da_j(\theta_i - b_j)]}{1 + \exp[Da_j(\theta_i - b_j)]}. \tag{5.2}$$

Other IRT models can be seen as special cases of the 3PL model. For instance, the two-parameter logistic IRT (2PL) model is obtained by setting $c_j = 0$ for all j, whereas the one-parameter logistic IRT (1PL) model additionally sets all a_j to be equal to 1. For details on IRT parameter estimation methods and software, the reader is referred to Fischer and Molenaar (1995), Baker and Kim (2004), and Tuerlinckx et al. (2004).

5.1.1 Scoring Using IRT Models

Using an IRT model to fit the binary $n_x \times J_x$ data matrix will produce both person and item parameter estimates $\hat{\theta}$ and $\hat{\omega}$ respectively. When an IRT model is used for scoring, instead of using the sum score $X_i \in \mathscr{X}$, an estimation of the ability parameter $\hat{\theta}$ is typically used as the test taker's score, and this is labelled the IRT score. In this case, $\hat{\theta} \in \Theta$, where Θ is the range of the IRT scores and is referred to as the IRT scale. A standard normal scale is often assumed, thus $\Theta \approx [-3, 3]$. It is, however, common to transform IRT scores into scale scores and to set a mean and standard deviation in order to avoid negative values for scores.

[1]The model shown in Example 1.2 corresponds to the *fixed-effects* version of IRT models. For more details on the difference between the fixed-effects version and the *random-effects* version of the model that is presented here, see San Martín et al. (2015).

5.2 Equating IRT Scores

To equate IRT scores under the IRT setting, we need a transformation function that maps the IRT scale on test form X to that of test form Y, i.e., $\varphi : \Theta_{\mathscr{X}} \mapsto \Theta_{\mathscr{Y}}$ (see, Definition 1.1). The transformation of IRT scales will be different depending on the equating design adopted. For instance, because in both the SG and the EG designs, the abilities are not a concern, no additional transformation of scales is needed as long as the estimation assumes the same constraints for the mean and variance of the ability distribution. When using a NEAT design, the groups are considered non-equivalent, so a transformation of the IRT scales is needed. In principle, only the transformation regarding θ is of interest for equating purposes (i.e. a transformation of IRT scales). Nevertheless, in the next section we introduce methods to transform the scale of all parameters involved (including both person and item parameters). Transforming item parameter scales is useful not only when IRT scales are linked, but also when number-correct scores are reported and equated instead of IRT scores. Other IRT equating methods that do not use a function to link IRT scales will be discussed in Sect. 5.4.

5.2.1 *Parameter Linking*

Because the parameters from different test forms need to be on the same scale, IRT *parameter linking* (von Davier and von Davier 2011) is conducted to place the IRT parameter estimates from separate calibrations of two test forms, on a common scale (Kolen and Brennan 2014). This is needed in particular when conducting equating under the NEAT design.

When an IRT model is used to fit two different test forms (see Kolen and Brennan 2014, Section 6.2), the transformation $\varphi : \Theta_{\mathscr{X}} \mapsto \Theta_{\mathscr{Y}}$ has typically been assumed to be a linear equation used to convert IRT scores as

$$\theta_{\mathscr{Y}i} = A\theta_{\mathscr{X}i} + B. \tag{5.3}$$

The relations between item parameters on the two test forms are as follows:

$$a_{\mathscr{Y}j} = a_{\mathscr{X}j}/A \tag{5.4}$$

$$b_{\mathscr{Y}j} = Ab_{\mathscr{X}j} + B \tag{5.5}$$

$$c_{\mathscr{Y}j} = c_{\mathscr{X}j}, \tag{5.6}$$

where A and B are linking constants (also known as equating coefficients) that are to be estimated. The indices $\mathscr{X}$ and $\mathscr{Y}$ are used to differentiate between the scales. A detailed account of the methods to calculate A and B can be found in Kolen and Brennan (2014, Chap. 6). A brief description of these methods is given in the following sections. These parameter linking methods are also used in Chap. 7 together with IRT kernel equating (Wiberg et al. 2014; Andersson and Wiberg 2016). A different perspective on item parameter linking establishing that "linking functions are not necessary to correct for arbitrary units and zeroes of the θ parameters but, more generally, to adjust for the different effects of the identifiability restrictions used in separate calibrations" is discussed in van der Linden and Barrett (2016) (see also, Bechger and Maris 2015).

5.2.1.1 Moments Methods to Estimate Equating Coefficients

These methods use different moments, i.e. the means and standard deviations of the common item parameter estimates, to obtain the equating coefficients A and B. Two methods were described in Marco (1977) and Loyd and Hoover (1980) and are refereed to in Kolen and Brennan (2014) as mean-mean and mean-sigma, respectively. Another method is referred to as mean-geometric mean (Mislevy and Bock 1990) and is described in, for example Ogasawara (2000). In all of these methods, the means and standard deviations are defined only on the set of common items between test forms X and Y. Let $\mu_{a\mathscr{X}}$ and $\mu_{a\mathscr{Y}}$ be the mean of the item discrimination parameter estimates taken only on the set of common items, and let $\sigma_{a\mathscr{X}}$ and $\sigma_{a\mathscr{Y}}$ be the corresponding standard deviations. The mean-mean method defines the equating coefficient A as

$$A = \frac{\mu_{a\mathscr{X}}}{\mu_{a\mathscr{Y}}}, \tag{5.7}$$

and the mean-sigma method defines the constant A as

$$A = \frac{\sigma_{b\mathscr{X}}}{\sigma_{b\mathscr{Y}}}. \tag{5.8}$$

The mean-geometric mean method defines the A constant as

$$A = \prod_{j=1}^{n_a} \left(\frac{a_{\mathscr{X}j}}{a_{\mathscr{Y}j}} \right)^{1/n_a}, \tag{5.9}$$

where n_a is the number of common items. Although different A:s are used in these three methods, the constant B is defined in each of these cases as

$$B = \mu_{b\mathscr{Y}} - A\mu_{b\mathscr{X}}. \tag{5.10}$$

5.2.1.2 Characteristic Curves Methods to Estimate Equating Coefficients

Two additional methods to obtain the equating coefficients were proposed by Haebara (1980) and Stocking and Lord (1983). These methods rely on the ICC:s and iteratively search for optimal A and B values by minimizing

$$\text{Hcrit} = \sum_i \sum_{j \in \mathscr{C}} \left[\pi_{ij}\left(\theta_{\mathscr{Y}i}, \hat{a}_{\mathscr{Y}j}, \hat{b}_{\mathscr{Y}j}, \hat{c}_{\mathscr{Y}j}\right) - \pi_{ij}\left(\theta_{\mathscr{Y}i}, \frac{\hat{a}_{\mathscr{X}j}}{A}, A\hat{b}_{\mathscr{X}j} + B, \hat{c}_{\mathscr{X}j}\right)\right]^2, \tag{5.11}$$

and

$$\text{SLcrit} = \sum_i \left[\sum_{j \in \mathscr{C}} \pi_{ij}\left(\theta_{\mathscr{Y}i}, \hat{a}_{\mathscr{Y}j}, \hat{b}_{\mathscr{Y}j}, \hat{c}_{\mathscr{Y}j}\right) - \sum_{j \in \mathscr{C}} \pi_{ij}\left(\theta_{\mathscr{Y}i}, \frac{\hat{a}_{\mathscr{X}j}}{A}, A\hat{b}_{\mathscr{X}j} + B, \hat{c}_{\mathscr{X}j}\right)\right]^2, \tag{5.12}$$

for the Haebara and Stocking-Lord methods, respectively. In both cases, $\mathscr{C}$ denotes the set of common items between test forms X and Y.

5.2.1.3 IRT Parameter Linking Using SNSequate

IRT parameter linking can be performed using the `irt.link()` function from the **SNSequate** package whose typical call reads as follows:

```
irt.link(parm, common, model, icc, D, ...)
```

The `irt.link()` function receives as arguments a data frame containing the item parameter estimates (`parm`), a numerical vector indicating the position where the common items are located (`comitems`), and both the type of the IRT model (`model`) and ICC (`icc`) that were used to obtain the item parameter estimates. It gives as output the values of the equating coefficients A and B calculated using the mean-mean, mean-sigma, Haebara, and Stocking-Lord methods.

Suppose we want to estimate the values of A and B that are needed to link the $\mathscr{X}$ scale to the $\mathscr{Y}$ scale (see Eqs. (5.3), (5.4), (5.5), and (5.6)). To illustrate how equating coefficients can be obtained using the function `irt.link()`, the `KB36` data object described in Chap. 2 is used. Remember that 12 out of the 36 items are common items between the test forms (i.e. every third item: 3, 6, 9, 12, 15, 18, 21, 24, 27, 30, 33, and 36). The following code can be used.

```
> library(SNSequate)
> data("KB36")
> parm.x = KB36$KBformX_par
> parm.y = KB36$KBformY_par
> comitems = seq(3, 36, 3)
> parm = as.data.frame(cbind(parm.y, parm.x))
```

```
> irt.link(parm, comitems, model = "3PL",
+ icc = "logistic", D = 1.7)
```

The `parm.x` and `parm.y` objects are three column matrices containing the item parameter estimates read from the `KB36` list. Because a 3PL model was fitted to obtain the item parameter estimates, the argument `model` was set to `"3PL"`. The corresponding output reads as

```
Call:
irt.link.default(parm = parm, common = comitems,
    model = "3PL", icc = "logistic", D = 1.7)

IRT parameter-linking constants:

                      A          B
Mean-Mean      1.217266 -0.5571557
Mean-Sigma     1.168891 -0.5155426
Haebara        1.093600 -0.4582959
Stocking-Lord  1.101954 -0.4770156
```

Note that the output reproduces the results reported in Table 6.6 in Kolen and Brennan (2014).

Although the logistic ICC is the most commonly used in IRT models for binary data, symmetric links are not always appropriate for modeling these kinds of data (Chen et al. 1999; Chen 2004). When asymmetric ICC:s are used in IRT models, methods of item parameter linking based on ICC:s (i.e., Haebara and Stocking-Lord) should accordingly be based on these asymmetric ICC:s (Estay 2012). This option is available in the `irt.link()` function when a 1PL model with asymmetric cloglog ICC is used to fit the data as illustrated in an example in González (2014).

5.2.1.4 IRT Parameter Linking Using equateIRT

IRT parameter linking is also implemented in the package **equateIRT** through the functions `modIRT()` and `direct()`. Typical calls to these functions are as follows.

```
modIRT(coef, var = NULL,...)

direc(mod1, mod2, method = "mean-mean",...)
```

The `coef` argument in the `modIRT()` function receives as input a matrix[2] containing the item parameter estimates. If the covariance matrix of item parameter estimates is available, **equateIRT** is also capable of estimating standard errors for the estimated equating coefficients. The argument `var` is used to read in the covariance matrix of item parameter estimates. The resulting `modIRT` object stores

[2]When multiple test forms are to be linked, the argument `coef` needs a list of matrices containing the item parameter estimates corresponding to each test form.

in a list the item parameter estimates and their covariance matrix (if supplied) for each of the tests forms to be linked.

Once a `modIRT` object has been created, it can be passed to the `direc()` function to perform the linking step itself. The arguments `mod1` and `mod2` are fed into the function using the elements of the list stored in the `modIRT` object.

It is important to emphasize that the row names in the matrix provided in the argument `coef` must be the names given to the items because this is the information that is used to differentiate between common and unique items in the linking process. This means that different names on the unique items and the same names on the common items are needed. For example, consider the `KB36` data where items $3, 6, 9, 12, 15, 18, 21, 24, 27, 30, 33,$ and 36 are common to both X and Y forms. If we decide to refer to the unique items in form X as `I1, I2, I4, I5, I7,...,I35` then different names for the non-common items in test form Y should be given. The following code shows a possible way to correctly assign names to the items.

```
> data("KB36",package="SNSequate")
> kbx<-cbind(KB36$KBformX_par[,3],KB36$KBformX_par[,2],
+ KB36$KBformX_par[,1])
> kby<-cbind(KB36$KBformY_par[,3],KB36$KBformY_par[,2],
+ KB36$KBformY_par[,1])

> row.names(kbx)<-paste("I", 1:36, sep = "")
> row.names(kby)<-paste("I", c(37,38, 3,39,40, 6,41,42,
+ 9,43,44, 12,45,46, 15,47,48, 18,49,50, 21,51,52,
+ 24,53,54, 27,55,56, 30,57,58, 33,59,60, 36), sep = "")

> datakb <-list(kbx,kby)
```

The element `kbx` in the list is a three column matrix with row names `I1`, `I2`, ..., `I36`. The `kby` element instead, appears as follows:

```
> datakb[[2]]
      [,1]    [,2]   [,3]
I37 0.1576 -1.4507 0.8704
I38 0.1094 -0.4070 0.4628
I3  0.1559 -1.3349 0.4416
I39 0.1381 -0.9017 0.5448
I40 0.2114 -1.4865 0.6200
I6  0.1913 -1.3210 0.5730
I41 0.2947  0.0691 1.1752
I42 0.2723  0.2324 0.4450
I9  0.1177 -0.7098 0.5987
I43 0.1445 -0.4253 0.8479
I44 0.0936 -0.8184 1.0320
I12 0.0818 -0.3539 0.6041
I45 0.1283 -0.0191 0.8297
I46 0.0854 -0.3155 0.7252
I15 0.3024  0.5320 0.9902
I47 0.2179  0.5394 0.7749
I48 0.2299  0.8987 0.5942
```

```
I18 0.0648 -0.1156 0.8081
I49 0.1633 -0.1948 0.9640
I50 0.1299  0.3506 0.7836
I21 0.2410  2.5538 0.4140
I51 0.1137 -0.1581 0.7618
I52 0.2397  0.5056 1.1959
I24 0.2243  0.5811 1.3554
I53 0.2577  0.6229 1.1869
I54 0.1856  0.3898 1.0296
I27 0.1651  0.9392 1.0417
I55 0.2323  1.1350 1.2055
I56 0.1070  0.6976 0.9697
I30 0.0794  1.8960 0.6336
I57 0.1855  1.3864 1.0822
I58 0.1027  0.9197 1.0195
I33 0.0630  1.0790 1.1347
I59 0.0999  1.8411 1.1948
I60 0.0832  2.0297 1.1961
I36 0.1259  2.1337 0.9255
```

where it can be seen that, indeed, the commons items have the same name as assigned in test form X.

Once the data have been properly prepared, the following code can be used to obtain the mean-mean equating coefficients.

```
> library(equateIRT)
> mod3plkb <- modIRT(coef = datakb,ltparam = FALSE,
+ lparam = FALSE)
> l12 <- direc(mod1 = mod3plkb[1], mod2 = mod3plkb[2],
+ method = "mean-mean")
> summary(l12)
Link: T1.T2
Method: mean-mean
Equating coefficients:
  Estimate StdErr
A  1.21727     NA
B -0.55716     NA
```

The `StdErr` column in the output shows `NA` because we did not provide the covariance matrix of the item parameter estimates. As mentioned before, when such a covariance matrix is available the program will provide standard errors for the equating coefficients A and B. To use any of the other methods to obtain the equating constants, we simply exchange the `method` argument with any of the following options: `mean-sigma`, `Stocking-Lord`, `Haebara`, or `mean-gmean`.

We have illustrated a simple linking procedure that considers only two test forms. The **equateIRT** package can, however, be used to implement more complex linkage plans that involve many tests forms. Also, we have used as input the item parameter estimates that were already available from the `KB36` data object. If this would not be the case, **equateIRT** can import item parameters estimates obtained from other packages such as **mirt** or **ltm** that are used to estimate IRT models. The binary data matrix can be modeled in either **ltm** or **mirt** and then the obtained results can be read into **equateIRT** using the `import.ltm()` and `import.mirt()` functions, respectively. This is exemplified using the **ltm** package to fit a 2PL model to both test forms' score data.

```
> library(ltm)
> modx.2pl<-ltm(KBneatX~z1)
> mody.2pl<-ltm(KBneatY~z1)
> estm2x <- import.ltm(modx.2pl,display=FALSE)
> estm2y <- import.ltm(mody.2pl,display=FALSE)
> pl2x <- modIRT(coef=list(estm2x$coef),
+ var=list(estm2x$var))
> pl2y <- modIRT(coef=list(estm2y$coef),
+ var=list(estm2y$var))
> pl2hs <- direc(mod1=pl2x[1], mod2 =pl2y[1],
+ method="Haebara")
> summary(pl2hs)
```

The output yields the equating coefficients for the Haebara method and appears as follows:

```
Link: T1.T1
Method: Haebara
Equating coefficients:
  Estimate   StdErr
A  0.88332 0.028071
B -0.44930 0.036127
```

The previous example can be replicated using **mirt** to obtain the corresponding item parameter estimates. Note that because different IRT software packages handle estimation differently, the results might differ slightly from analyses with other IRT software packages.

5.3 Equating Observed Scores Under the IRT Framework

Even when test forms are scored using IRT scores (i.e., using $\hat{\theta}_i$ as a test-taker's i score), it is still possible to report results under the observed score scale using the sum scores $X_i = \sum_{j=1}^{J_x} X_{ij}$ and $Y_i = \sum_{j=1}^{J_y} Y_{ij}$, which are obtained from the binary data matrix. Although in principle any of the equating transformations described in the previous chapters could be used to map $\mathscr{X}$ onto $\mathscr{Y}$, two model-theoretic methods based on IRT have been widely used for this purpose. These methods are called IRT true-score equating and IRT observed-score equating (Lord 1980; Lord and Wingersky 1984) and are described next. Other IRT equating methods are discussed in Chaps. 6 and 7. Both methods rely on the comparison of the score distributions, but a significant difference with the methods described in the previous chapters is that both are based on conditional distributions of tests scores given the ability (González et al. 2016). Using IRT in the process of equating sum scores, rather than IRT scores, requires item parameter linking as a preliminary step. After item parameters are estimated and placed on the same scale using some of the methods described in Sect. 5.2.1, IRT equating transformations of the type to be defined in Sects. 5.3.1 and 5.3.2 are used to relate scores from two test forms in the usual way. In the exposition that follows, we assume that $\mathbf{X} = (X_1, \ldots, X_{J_x})$ is the response pattern for a test taker of ability θ such that the corresponding sum score is $X = \sum_{j=1}^{J_x} X_{ij}$.

5.3.1 *IRT True-Score Equating*

IRT true-score equating (Lord 1980) is based on the mean of the conditional score distribution. In classical test theory (Lord and Novick 1968), a true score is defined as the conditional expectation of the score random variable given the test taker's ability. If τ_X and τ_Y represent the true scores associated with test forms X and Y, then

$$\tau_X = E(X \mid \theta) = \sum_{j:\mathscr{X}} \pi_{ij}(\theta, \omega_j) = T^{\tau_X}(\theta), \tag{5.13}$$

$$\tau_Y = E(Y \mid \theta) = \sum_{j:\mathscr{Y}} \pi_{ij}(\theta, \omega_j) = T^{\tau_Y}(\theta), \tag{5.14}$$

where the sum is over the items of the corresponding test form. In the psychometric literature, the $T(\cdot)$ functions are known as test characteristic functions (Lord 1980).

Equations (5.13) and (5.14) are used to relate true scores τ_X and τ_Y. Thus, a true score τ_X associated with a given θ is considered to be equivalent to the true score τ_Y that is associated with that same θ. The resulting transformation to find an equivalent true score τ_Y of τ_X is given by

$$\varphi(\tau_X, \theta) = T^{\tau_Y}(T^{\tau_X^{-1}}(\tau_X)). \tag{5.15}$$

In practice, estimates of the item parameters for each test form are used to produce an *estimated true score relationship* $\hat{\varphi}(\tau_X, \theta, \hat{\omega}_{\mathscr{X}}, \hat{\omega}_{\mathscr{Y}})$ by means of the test characteristics functions. Because true scores are not observable, the estimated true score conversion is actually applied to the observed sum scores X and Y. Thus, in practice, the method starts by solving for θ in Eq. (5.13) for a given value of x. Then, the obtained value of θ is plugged into Eq. (5.14) to obtain the corresponding equated value y. An equating transformation based not only on the mean but on the whole conditional score distributions will be described in Chap. 6 for the local equating method.

5.3.2 *IRT Observed-Score Equating*

IRT observed-score equating (Lord 1980; Lord and Wingersky 1984) is based on the marginal score distributions and uses the IRT model to define the conditional score probabilities involved. To obtain a marginal score distribution, a distribution for the abilities is assumed and integrated (or summed) across all ability levels to produce a marginal observed score distribution. Once this process has been completed for both X and Y, equipercentile equating is applied to relate scores between the two test forms. Formally, let $f(x \mid \theta), f(y \mid \theta)$, and $\psi(\theta)$ be the conditional scores and ability distributions, respectively, then the resulting equating transformation is

$$\varphi(x) = F_Y^{-1}(F_X(x)), \tag{5.16}$$

where F_X and F_Y are obtained after cumulating the values of the two mixtures $f(x) = \int f(x \mid \theta)\psi(\theta)d\theta$ and $f(y) = \int f(y \mid \theta)\psi(\theta)d\theta$. To obtain the conditional scores distributions, a recursive algorithm proposed in Lord and Wingersky (1984) is most commonly used, although new alternatives have been proposed recently (González et al. 2016). Details of the Lord and Wingersky algorithm are given in Sect. B.5.

5.3.3 *IRT True-Score and Observed-Score Equating Using SNSequate*

The `irt.eq()` function in the **SNSequate** package implements both IRT true-score and IRT observed-score equating. The general function call is

```
irt.eq(n_items, param_x, param_y, theta_points=NULL,
weights=NULL, n_points=10, w=1, A=NULL, B=NULL,
link=NULL, method_link=NULL, common=NULL, method="TS",
D=1.7,...)
```

The function receives as its main arguments the number of items in the test, `n_items`, and two sets of item parameter estimates, $\hat{\omega}_{\mathscr{X}}$ and $\hat{\omega}_{\mathscr{Y}}$, which are labelled `param_x` and `param_y`, respectively. The default method of IRT equating is true-score equating and it is obtained by setting the argument `method="TS"`. When IRT observed-score is used (`method="OS"`), two additional arguments need to be included. The `theta_points` and `weights` arguments are used to build the ability distribution using a grid of possible values of θ to integrate out the ability term. If `weights=NULL`, the method assumes that the distribution of ability is characterized by a finite number of abilities (Kolen and Brennan 2014, p.199). The equating coefficients A and B can either be provided by the user using the arguments `A` and `B`, respectively, or they can be automatically[3] calculated providing the set of common items in the argument `common` and the preferred linking method in the argument `method_link` (e.g., `mean/mean`, `mean/sigma`, `Haebara`, `StockLord`). It is also possible to use a previously created `irt.link` object as an input in the argument `link`. The function gives as an output the table of converted scores.

[3]In this case, an internal call to `irt.link()` is made.

To illustrate the use of `irt.eq()`, we again use the `KB36` data from Kolen and Brennan (2014). The `KB36_t` object contains all the information that appears in Table 6.8 in Kolen and Brennan (2014). We can obtain the item parameter estimates using the following code

```
> data("KB36_t", package="SNSequate")
> param_x <- list(a=KB36_t[,3],b=KB36_t[,4],c=KB36_t[,5])
> param_y <- list(a=KB36_t[,7],b=KB36_t[,8],c=KB36_t[,9])
```

In order to obtain comparable results for the IRT observed-score method, we use the same points and weights for θ as the ones shown in Table 6.10 in Kolen and Brennan (2014)

```
> theta_points=c(-5.2086,-4.163,-3.1175,-2.072,-1.0269,
                 0.0184,1.0635,2.109,3.1546,4.2001)
> weights=c(0.000101,0.00276,0.03021,0.142,0.3149,0.3158,
            0.1542,0.03596,0.003925,0.000186)
```

Finally, the equating is performed. The following code[4] can be used to obtain a table of transformed scores using both IRT true-score equating (IRTTSE) and IRT observed-score equating (IRTOSE).

```
> res.tse<-irt.eq(36, param_x, param_y, method="TS",
+ A=1, B=0)
> res.ose<-irt.eq(36, param_x, param_y, theta_points,
+ A=1, B=0, weights, method="OS")
> outirt<-cbind(Theta=res.tse$theta_equivalent,
+ Scale=0:36,IRTTSE=res.tse$tau_y,IRTOSE=res.ose$e_Y_x)
```

The obtained output is as follows.

```
> outirt
            Theta Scale     IRTTSE      IRTOSE
 [1,]          NA     0  0.0000000  0.03504169
 [2,]          NA     1  0.8880207  0.61791013
 [3,]          NA     2  1.7760414  1.58016093
 [4,]          NA     3  2.6640621  2.54598504
 [5,]          NA     4  3.5520829  3.51853882
 [6,]          NA     5  4.4401036  4.50262990
 [7,]          NA     6  5.3281243  5.50482544
 [8,] -4.33579405     7  6.1340702  6.53178001
 [9,] -2.76962168     8  7.1863245  7.58600535
[10,] -2.06273325     9  8.3961646  8.66191273
[11,] -1.60659143    10  9.6235960  9.74810798
[12,] -1.26759951    11 10.8278915 10.83647390
[13,] -0.99453316    12 12.0027423 11.93041351
[14,] -0.76275510    13 13.1523676 13.04561220
[15,] -0.55875957    14 14.2832934 14.19714790
```

[4]Note that the item parameter estimates shown in Table 6.10 in Kolen and Brennan (2014) are already rescaled. This is why we have set the equating coefficients as `A=1` and `B=0` so that comparable results with those obtained in Kolen and Brennan (2014) are obtained.

```
[16,] -0.37426356      15 15.4026505 15.37001929
[17,] -0.20386899      16 16.5163433 16.51360703
[18,] -0.04361758      17 17.6301154 17.59791730
[19,]  0.10909751      18 18.7454840 18.64431311
[20,]  0.25645197      19 19.8633271 19.67923022
[21,]  0.40000362      20 20.9809800 20.73895536
[22,]  0.54100388      21 22.0936749 21.87754232
[23,]  0.68054901      22 23.1955335 23.10284297
[24,]  0.81968954      23 24.2805929 24.28949855
[25,]  0.95967002      24 25.3448289 25.36173053
[26,]  1.10197542      25 26.3862698 26.36432784
[27,]  1.24876542      26 27.4072685 27.34317400
[28,]  1.40277618      27 28.4119269 28.32171394
[29,]  1.56770324      28 29.4060002 29.31912794
[30,]  1.74866222      29 30.3953420 30.35038228
[31,]  1.95274485      30 31.3819416 31.37659605
[32,]  2.19095382      31 32.3613840 32.34521624
[33,]  2.48169050      32 33.3158694 33.28005241
[34,]  2.85959583      33 34.2081257 34.19868532
[35,]  3.39838079      34 34.9791161 35.07472658
[36,]  4.32073754      35 35.5753858 35.85184549
[37,]          NA      36 36.0000000 36.39026070
```

A graphical way to compare both methods is by plotting the estimated equating relationships for IRT true-score and IRT observed-score equating. Figure 5.1 shows a plot where test form X scores are subtracted from equated scores and plotted against the scale scores.

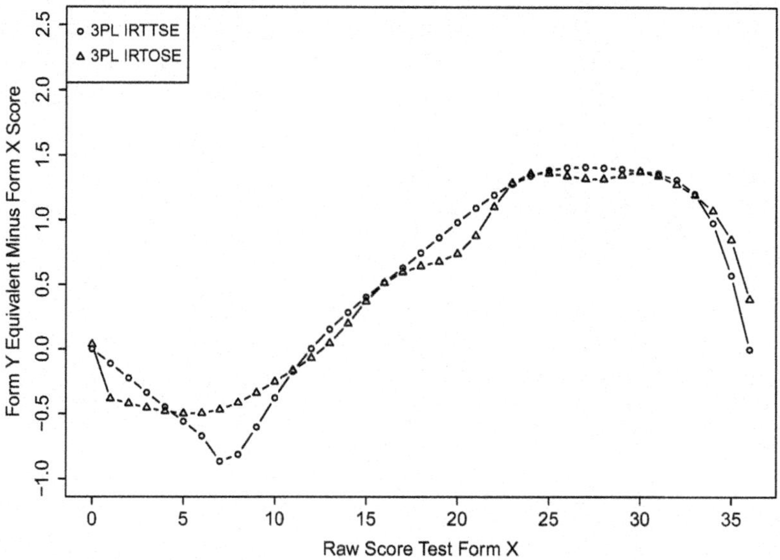

Fig. 5.1 IRT true-score and IRT observed-score equating example from Kolen and Brennan (2014)

```
> plot(0:36,outirt[,3]-0:36,ylim=c(-1,2.5),type='b',
+ pch=1,ylab='Form Y Equivalent Minus Form X Score',
+ xlab='Raw Score Test Form X')
> points(0:36,outirt[,4]-0:36,type='b',pch=2)
> legend('topleft',pch=c(1,2),c("3PL IRTTSE",
+ "3PL IRTOSE"))
```

Note that the figure intends to reproduce what is shown in Figure 6.6 in Kolen and Brennan (2014).[5]

As mentioned before, IRT equating requires the use of item parameter estimates obtained from an IRT software package. In the following example, we perform a Rasch equating, mimicking the illustration in Sect. 6.8.4 in Kolen and Brennan (2014). To illustrate the fitting of IRT models, we use the **ltm** package. The item parameter estimates from the obtained output are then used in **SNSequate** to perform IRT true-score and IRT observed-score equating using the Rasch model. First, the IRT models are fitted

```
> library(ltm)
> modx.rasch <-rasch(KBneatX, constraint =
+ cbind(ncol(KBneatX) + 1, 1))
> mody.rasch <-rasch(KBneatY, constraint =
+ cbind(ncol(KBneatY) + 1, 1))
```

The `rasch()` function is used to fit the Rasch model under the IRT framework. It receives as input the binary data matrix of item responses (in this case, the object `KBneatX`) and returns as its main output the item difficulty parameter estimates. The argument `constraint` is used to force the value of all discrimination parameters, $a_j, j = 1, \ldots, J_x$, to be equal to 1. Otherwise, a common value of a for all items in the test will be estimated. Item parameter estimates can be retrieved from a `ltm` object using the function `coef()`. The full sets of item parameter estimates are stored in the objects `param_x` and `param_y`, respectively.[6] The following code is used to perform these tasks

```
> parm.x = coef(modx.rasch)[,1]
> parm.y = coef(mody.rasch)[,1]
> param_x <- list(a=rep(1,36),b=parm.x,c=rep(0,36))
> param_y <- list(a=rep(1,36),b=parm.y,c=rep(0,36))
```

The actual Rasch equating (IRT true-score and IRT observed-score equating) is carried out using the `irt.eq()` function. The following code shows how to implement both IRT true-score and IRT observed-score equating.

```
> res.Rasch.tse<-irt.eq(36, param_x, param_y,
+ method="TS",method_link="mean/sigma", common=seq(3, 36, 3))
```

[5]Figure 6.6 in Kolen and Brennan (2014) also shows the curve for frequency estimation equating. This curve can easily be obtained and added using the **equate** package as illustrated in Chap. 3.

[6]Because a Rasch model is used to fit the 0/1 data, item discrimination parameters are fixed to 1 and guessing parameters fixed to 0.

```
> res.Rasch.ose<-irt.eq(36, param_x, param_y,
+ method="OS",method_link="mean/sigma", common=seq(3, 36, 3))
> outirt.Rasch <- cbind(Theta=res.Rasch.tse$theta_equivalent,
+ Scale=0:36,IRTTSE=res.Rasch.tse$tau_y, IRTOSE=res.Rasch.ose
  $e_Y_x)

> outirt.Rasch
            Theta Scale    IRTTSE     IRTOSE
 [1,]          NA     0  0.000000  0.1467132
 [2,] -2.80749301     1  1.045037  1.1022981
 [3,] -2.31256342     2  2.147447  2.2394178
 [4,] -1.99811265     3  3.256670  3.4056409
 [5,] -1.76024710     4  4.356351  4.4753960
 [6,] -1.56518154     5  5.443147  5.4884525
 [7,] -1.39728413     6  6.518522  6.4916573
 [8,] -1.24805199     7  7.584381  7.5065491
 [9,] -1.11235774     8  8.641921  8.5863401
[10,] -0.98666501     9  9.693277  9.7274636
[11,] -0.86859558    10 10.739184 10.8125062
[12,] -0.75637361    11 11.780276 11.8436280
[13,] -0.64867721    12 12.816507 12.8576140
[14,] -0.54442483    13 13.847948 13.8664930
[15,] -0.44272917    14 14.874489 14.8729593
[16,] -0.34281935    15 15.896081 15.8776476
[17,] -0.24407622    16 16.911997 16.8811194
[18,] -0.14591665    17 17.921735 17.8853089
[19,] -0.04771696    18 18.925594 18.8940060
[20,]  0.05095184    19 19.922097 19.9010902
[21,]  0.15064579    20 20.911129 20.8949148
[22,]  0.25187582    21 21.892041 21.8800825
[23,]  0.35522477    22 22.864652 22.8605112
[24,]  0.46122338    23 23.827922 23.8363160
[25,]  0.57058510    24 24.781942 24.8060091
[26,]  0.68407215    25 25.726411 25.7659877
[27,]  0.80260058    26 26.661245 26.7068620
[28,]  0.92733269    27 27.586818 27.6137969
[29,]  1.05963287    28 28.502990 28.5096457
[30,]  1.20167135    29 29.412190 29.4049402
[31,]  1.35620113    30 30.315793 30.3052847
[32,]  1.52769604    31 31.217647 31.2295758
[33,]  1.72350616    32 32.122877 32.1670270
[34,]  1.95764962    33 33.039835 33.1029061
[35,]  2.26159884    34 33.980135 34.0231544
[36,]  2.73617518    35 34.960572 34.9650898
[37,]          NA    36 36.000000 35.9597041
```

Note that the output mimics what is shown in Table 6.14 in Kolen and Brennan (2014). It should be mentioned however, that results are not expected to be exactly the same as the ones shown in Table 6.14 in Kolen and Brennan (2014) because (i) different IRT software packages with possible different default settings were used (i.e., **BILOG-MG** vs. **ltm**), (ii) different equating coefficients could have been used for item parameter linking (i.e., it is not specified in Kolen and Brennan

(2014) which IRT linking method was used to obtain the equating coefficients), and (iii) different ways to average the ability distribution could have been used (i.e., for simplicity we used the same points and weights as the ones used for the 3PL example, but it is not specified in Kolen and Brennan (2014) how the ability distribution was integrated out).

To end this section, we continue the previous Rasch IRT true-score and IRT observed-score equating example above by illustrating how to use **equateIRT** to link item difficulty parameters using the Stocking-Lord method.

```
> est.modx.rasch <- import.ltm(modx.rasch)
> est.mody.rasch <- import.ltm(mody.rasch)

> modrasch.ltm <- modIRT(coef = list(est.modx.rasch$coef,
+ est.mody.rasch$coef), var = list(est.modx.rasch$var,
+ est.mody.rasch$var), digits = 4)

> l12.ltm <- direc(mod1 = modrasch.ltm[1],
+ mod2 = modrasch.ltm[2], method = "Stocking-Lord")
> conv<-convert(A = l12.ltm$A, B = l12.ltm$B, coef =
+ coef(modrasch.ltm$T1))
> conv$coef
  Dffclt.It1   Dffclt.It2   Dffclt.It3   Dffclt.It4
-2.370672800 -1.255691866 -1.445008169 -0.637413454
  Dffclt.It5   Dffclt.It6   Dffclt.It7   Dffclt.It8
-1.283871507 -1.664591169 -0.870807837 -1.156791618
  Dffclt.It9  Dffclt.It10  Dffclt.It11  Dffclt.It12
-0.832791526 -0.586065535 -0.363761470 -0.717535419
 Dffclt.It13  Dffclt.It14  Dffclt.It15  Dffclt.It16
-0.640265086  0.007582455 -0.338004506 -0.512003500
 Dffclt.It17  Dffclt.It18  Dffclt.It19  Dffclt.It20
-0.529100461 -0.363757637 -0.403735303 -0.007356718
 Dffclt.It21  Dffclt.It22  Dffclt.It23  Dffclt.It24
 0.334935640 -0.113030497  0.112082500  0.001493633
 Dffclt.It25  Dffclt.It26  Dffclt.It27  Dffclt.It28
 0.179039907  0.511109405  0.578115877  0.779462425
 Dffclt.It29  Dffclt.It30  Dffclt.It31  Dffclt.It32
 0.711007422  1.316157527  0.521125051  1.079218991
 Dffclt.It33  Dffclt.It34  Dffclt.It35  Dffclt.It36
 1.311846421  1.321117518  1.780480861  1.554429979
```

The obtained values can be compared to those shown in Table 6.13 in Kolen and Brennan (2014).

5.3.4 *IRT True-Score and Observed-Score Equating Using equateIRT*

The `score()` function in the **equateIRT** package can be used to obtain equated scores using either IRT true-score or IRT observed-score equating. We exemplify the use of `score()` with the ADM data. In this example, we illustrate first how to fit a 2PL IRT model to the binary data using the **mirt** package, and then how parameter estimates can be read into **equateIRT**. Next, item parameter linking is performed prior to the IRT observed-score equating. The following code shows how to implement all of these steps.

```
> load(url("http://www.mat.uc.cl/~jorge.gonzalez/
+ EquatingRbook/ADMneatX.Rda"))
> load(url("http://www.mat.uc.cl/~jorge.gonzalez/
+ EquatingRbook/ADMneatY.Rda"))
> library(mirt)
> modADMx.2PL <- mirt(ADMneatX, 1, itemtype = "2PL",
+ SE = TRUE, SE.type = 'BL')
> modADMy.2PL <- mirt(ADMneatY, 1, itemtype = "2PL",
+ SE = TRUE, SE.type = 'BL')
```

The objects `modADMx.2PL` and `modADMy.2PL` store the 2PL fit to the response data from test forms X and Y, respectively. The obtained item parameter estimates can be imported to **equateIRT** using the `import.mirt()` function as follows.

```
> parADMx.2PL<-import.mirt(modADMx.2PL, display = FALSE,
+ digits = 3)
> parADMy.2PL<-import.mirt(modADMy.2PL, display = FALSE,
+ digits = 3)
```

In Sect. 5.2.1.4, we saw that common items must share the same names in order to be distinguished from unique test form items. In the ADM data, the first 40 (out of 120) items in each test form correspond to anchor items. We give the names `c1`, ... ,`c40` to the common items and the names `uX` and `uY` for the unique items in test forms X and Y, respectively. The item parameter estimates with the relabeled item names are then stored in the list `parADM.2PL` as shown in the following code.

```
> aux.x<-as.matrix(parADMx.2PL$coef)
> aux.y<-as.matrix(parADMy.2PL$coef)
> row.names(aux.x)<-c(paste("c", 1:40, sep = ""),
+ paste("uX", 1:80, sep = ""))
> row.names(aux.y)<-c(paste("c", 1:40, sep = ""),
+ paste("uY", 1:80, sep = ""))
> parADM.2PL<-list(aux.x,aux.y)
```

Before the equating step, item parameter linking is conducted in order to put the item parameter estimates on a common scale. In this case, we use the mean-sigma method to obtain equating coefficients.

```
> mod2pl.adm <- modIRT(coef = parADM.2PL,
+ ltparam = FALSE, lparam = FALSE)
> l12 <- direc(mod1 = mod2pl.adm[1],
+ mod2 = mod2pl.adm[2], method = "mean-sigma")
```

The final step is the equating itself, which is performed using the `score()` function as shown below.

```
> score(l12, method = "OSE",se=FALSE,scores = 5:15)
   T2   T1.as.T2
6   5   5.078680
7   6   6.098008
8   7   7.121798
9   8   8.150029
10  9   9.182403
11 10  10.218250
12 11  11.256572
13 12  12.296327
14 13  13.336815
15 14  14.377856
16 15  15.419550
```

The argument `scores` indicates that only the equated values for scores 5 to 15 in the scale are to be returned. If this argument is not set to a particular set of values, then the complete list of equated values is returned.

5.4 Other Equating Methods for IRT Scores

In Sect. 5.2, IRT scores were linked using a linear function $\varphi : \Theta_{\mathscr{X}} \mapsto \Theta_{\mathscr{Y}}$ defined by the equating coefficients A and B (see Eq. (5.3)). The linking function in Eq. (5.3) can be seen as an equating transformation that maps IRT scores from one IRT scale onto the other. In the following sections, we describe two alternative methods to link IRT scores known as concurrent calibration and fixed item parameter calibration. A notable difference is that these alternative methods do not need an equating transformation to perform the equating, and instead parameters are linked on a common scale during the estimation routine.

5.4.1 Concurrent Calibration

In the *concurrent calibration* method (Wingersky and Lord 1984), the parameters obtained from both test forms data (i.e., $\omega_{\mathscr{X}}$ and $\omega_{\mathscr{Y}}$) are estimated together in a single run of the IRT software where separate ability distributions are assumed in the two populations. Items that are not common are treated as *not reached* by the program. As an example, suppose that two test forms X and Y are composed of six items in total, where three of them are unique items and three are common items. Also, assume that a sample from population P is composed of three test takers whereas in population Q the sample size is four test takers. A schematic representation of the data structure necessary to perform concurrent calibration in this case is shown in Fig. 5.2. As mentioned before, because the two test taker

		X			A			Y		
	item id	it1	it2	it3	a1	a2	a3	it1	it2	it3
P	1	✓	✓	✓	✓	✓	✓	·	·	·
	2	✓	✓	✓	✓	✓	✓	·	·	·
	3	✓	✓	✓	✓	✓	✓	·	·	·
Q	1	·	·	·	✓	✓	✓	✓	✓	✓
	2	·	·	·	✓	✓	✓	✓	✓	✓
	3	·	·	·	✓	✓	✓	✓	✓	✓
	4	·	·	·	✓	✓	✓	✓	✓	✓

Fig. 5.2 Schematic representation of the data structure to perform concurrent calibration

populations could differ significantly in ability, the IRT software used for concurrent calibration must have a feature that supports multiple groups for the estimation. This means that a different distribution for abilities is assumed for each of the groups in the calibration. DeMars (2002) pointed out that item difficulty parameters can be overestimated for the items on the less difficult test form and underestimated for the items on the more difficult test form if group differences in ability are not taken into account. However, the simulation study in DeMars (2002) showed that the amount of bias was small (at least for the studied conditions) such that large practical differences are not expected. Based on this last finding, in the following section we give an example of how to perform concurrent calibration using the **ltm** package, which does not support the multiple group feature.

5.4.1.1 Concurrent Calibration Using ltm

The following example shows how to implement concurrent calibration when the **ltm** package is used to fit a Rasch model, using the `KB` data. First, we need to create a data set with the structure shown in Fig. 5.2. The code below perform this task.

```
> data("KB36", package = "SNSequate")
> KBneatX<-KB36$KBformX
> KBneatY<-KB36$KBformY
> # extracting common items
> xci<-KBneatX[,seq(3,36,3)]
> yci<-KBneatY[,seq(3,36,3)]

> xui<-KBneatX[,setdiff(seq(1:36),seq(3,36,3))]
> yui<-KBneatY[,setdiff(seq(1:36),seq(3,36,3))]

> nax<-matrix(NA,nrow=1655,ncol=24)
> nay<-matrix(NA,nrow=1638,ncol=24)

> colnames(nax)<-paste("It",seq(37,60,1),sep="")
> colnames(nay)<-paste("It",setdiff(seq(1:36),
+ seq(3,36,3)),sep="")
> colnames(yui)<-paste("It",seq(37,60,1),sep="")
```

```
> kbfx<-cbind(xui,xci,nax)
> kbfy<-cbind(nay,yci,yui)

> # final data to be concurrently calibrated
> datkb<-rbind(kbfx,kbfy)
```

The object `datkb` is a 3293×60 matrix. The total number of rows is the sum of the number of test takers for both X and Y ($1655 + 1638 = 3293$). The total number of columns corresponds to the sum of the unique items for X and Y and the common items ($24 + 24 + 12 = 60$). The *not reached* items are represented by `NA`. Using this data set, the Rasch concurrent calibration can be stored in the object `modRasch.cc` and can be performed as follows.

```
> modRasch.cc<-rasch(datkb, constraint =
+ cbind(ncol(datkb) + 1, 1))
```

For comparison purposes, we also fit separate calibrations stored in the objects `modx.rasch` and `mody.rasch` using test forms X and Y, respectively. Because IRT scores are being equated, a graphical comparison can be made by plotting the predicted ability distributions using the results from the concurrent and the separate calibrations. This plot is shown in Fig. 5.3. The **R** code used for the separate calibrations as well as to produce the plot of the predicted ability distributions is shown below.

```
> modx.rasch<-rasch(KBneatX, constraint =
+ cbind(ncol(KBneatX) + 1, 1))
> mody.rasch<-rasch(KBneatY, constraint =
+ cbind(ncol(KBneatY) + 1, 1))
> coef(modx.rasch)
> coef(mody.rasch)
> coef(modRasch.cc)
```

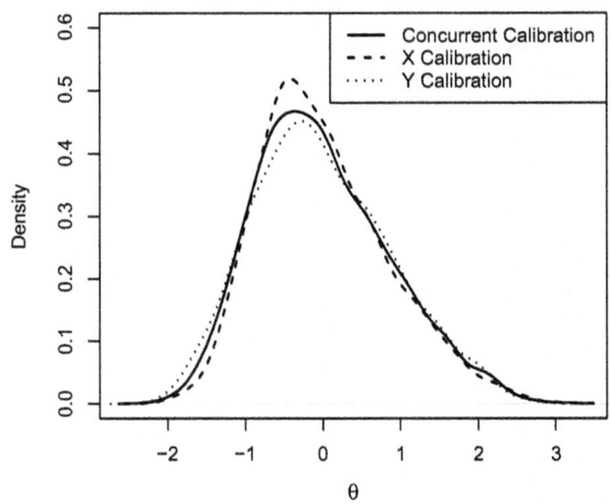

Fig. 5.3 Predicted ability distributions for separate calibration of both test forms and the concurrent calibration

```
> thetas.xr<-factor.scores(modx.rasch)
> thetas.yr<-factor.scores(mody.rasch)
> thetas.cc<-factor.scores(modRasch.cc)

> plot(density(thetas.cc$score.dat$z1),ylim=c(0,0.6),
+ main="",xlab=expression(theta),lwd=2.0,lty=1)
> lines(density(thetas.xr$score.dat$z1),lwd=2.0,lty=2)
> lines(density(thetas.yr$score.dat$z1),lwd=2.0,lty=3)
> legend("topright",lty=c(1,2,3),lwd=c(2,2,2),
+ c("Concurrent Calibration","X Calibration",
+ "Y Calibration"))
```

5.4.2 *Fixed Item Parameter Calibration*

In the *fixed item parameter calibration* method (Kim 2006), the estimated values for common items from previous calibrations are fixed in the calibration of other test forms. Assuming that test form Y is first administered, parameter estimates for the portion of common items are used and treated as fixed when calibrating test form X. In this way, the $\mathscr{X}$ scale is forced to be on the $\mathscr{Y}$ scale.

5.4.2.1 Fixed Item Parameter Calibration Using mirt

We use the `KB36` data to show how fixed item parameter calibration can be performed using the **mirt** package. A key task is to instruct the program to fix the values of anchor parameter estimates when calibrating a new test form. This task can be performed using the argument `parms` in the `mirt()` function. By setting `parms="values"`, it is possible to inspect (and later modify) the way the model is being estimated (i.e., which starting values are being used for estimation, which parameter are actually being estimated and which are left fixed, etc.). Suppose that test form Y is first calibrated. The previously explained steps can be performed using the following code.[7]

```
> str.mod <- mirt(KB36$KBformY, 1, itemtype = 'Rasch',
+ pars = 'values')
> head(str.mod,n=12)

   group item class name parnum     value lbound ubound   est
1    all  It1  dich   a1      1 1.0000000   -Inf    Inf FALSE
2    all  It1  dich    d      2 2.0612449   -Inf    Inf  TRUE
3    all  It1  dich    g      3 0.0000000      0      1 FALSE
4    all  It1  dich    u      4 1.0000000      0      1 FALSE
5    all  It2  dich   a1      5 1.0000000   -Inf    Inf FALSE
6    all  It2  dich    d      6 0.5828647   -Inf    Inf  TRUE
```

[7]Some columns in the output are omitted.

```
7     all  It2  dich    g      7 0.0000000     0       1 FALSE
8     all  It2  dich    u      8 1.0000000     0       1 FALSE
9     all  It3  dich   a1      9 1.0000000  -Inf     Inf FALSE
10    all  It3  dich    d     10 1.3540319  -Inf     Inf  TRUE
11    all  It3  dich    g     11 0.0000000     0       1 FALSE
12    all  It3  dich    u     12 1.0000000     0       1 FALSE
```

The column `est` in the output indicates that the only parameter[8] that is being estimated is `d` (see column `name` in the output). Because in the `KB36` data every three items are common items, the `est` column can be modified in the following way.

```
> str.mod$est <- c(rep(c(FALSE,TRUE,FALSE,FALSE,
                    + FALSE,TRUE,FALSE,FALSE,
                    + FALSE,FALSE,FALSE,FALSE),12),FALSE,
                    + FALSE)
```

This modification will result in anchor items being fixed to the values estimated in the calibration of test form X, and unique items in test form X will indeed be estimated. The calibration of test form X using fixed item anchor items is performed as follows.

```
> mod.FIPC <- mirt(KB36$KBformX, 1, pars = str.mod)
```

In order to check that anchor items were indeed fixed and that unique items were estimated, we also calibrate test forms X and Y separately and compare the resulting item parameter estimates with the fixed item parameter calibration results.

```
> mod.x<-mirt(KB36$KBformX, 1, itemtype = 'Rasch')
> mod.y<-mirt(KB36$KBformY, 1, itemtype = 'Rasch')

# Check that anchor items were fixed
> aux.parm<-matrix(c(str.mod$val,0,0),ncol=4,byrow=TRUE)
> check.parm<-cbind(aux.parm[-37,],coef(mod.FIPC,
+ simplify=TRUE)$items)
> check.parm
                          a1     d g u
It1  1  2.061244896 0 1  1  2.176 0 1
It2  1  0.582864689 0 1  1  1.062 0 1
It3  1  1.354031884 0 1  1  1.354 0 1
It4  1  1.115791050 0 1  1  0.444 0 1
It5  1  1.872662531 0 1  1  1.090 0 1
It6  1  1.618311781 0 1  1  1.618 0 1
It7  1  0.681053078 0 1  1  0.677 0 1
It8  1  0.548382925 0 1  1  0.963 0 1
It9  1  0.950015430 0 1  1  0.950 0 1
It10 1  0.836619693 0 1  1  0.392 0 1
```

[8]The `mirt()` function implement a general four parameter model from which the 1PL, 2PL and 3PL models are particular cases. The discrimination, difficulty and guessing parameters are denoted by `a1`, `d`, and `g`, respectively, whereas a fourth upper asymptote parameter is denoted by `u`. In the case of the Rasch model, `a1=u=1` and `c=0`.

```
It11 1  1.289923157 0 1  1  0.170 0 1
It12 1  0.539008477 0 1  1  0.539 0 1
It13 1  0.341558648 0 1  1  0.446 0 1
It14 1  0.548382925 0 1  1 -0.201 0 1
It15 1  0.265464639 0 1  1  0.265 0 1
It16 1  0.048124734 0 1  1  0.318 0 1
It17 1 -0.102301021 0 1  1  0.335 0 1
It18 1  0.292806712 0 1  1  0.293 0 1
It19 1  0.636514043 0 1  1  0.210 0 1
It20 1 -0.042108156 0 1  1 -0.187 0 1
It21 1 -0.671480291 0 1  1 -0.671 0 1
It22 1  0.452072063 0 1  1 -0.081 0 1
It23 1  0.048124734 0 1  1 -0.306 0 1
It24 1 -0.108324656 0 1  1 -0.108 0 1
It25 1 -0.009022518 0 1  1 -0.373 0 1
It26 1  0.030076127 0 1  1 -0.705 0 1
It27 1 -0.611210637 0 1  1 -0.611 0 1
It28 1 -0.570305655 0 1  1 -0.974 0 1
It29 1 -0.548382925 0 1  1 -0.905 0 1
It30 1 -1.490825665 0 1  1 -1.491 0 1
It31 1 -0.933149123 0 1  1 -0.715 0 1
It32 1 -0.830038920 0 1  1 -1.274 0 1
It33 1 -1.238142370 0 1  1 -1.238 0 1
It34 1 -1.821213392 0 1  1 -1.516 0 1
It35 1 -2.092787071 0 1  1 -1.976 0 1
It36 1 -1.682497427 0 1  1 -1.682 0 1
```

The output shows that, indeed, for every three items the values of the parameter estimates are the same for both test forms.

5.5 Other R Packages for IRT Analysis

In this chapter, we have illustrated how the **ltm** and **mirt** packages can be used to fit IRT models and how the obtained results from these packages can be combined with **SNSequate** and **equateIRT** for performing equating tasks. There are also other **R** packages that can be used for different types of IRT analyses. For example, the `lmer()` function in the **lme4** package (Bates et al. 2015) can be used to fit several IRT models (De Boeck et al. 2011). The `sca()` function in the **irtoys** package (Partchev 2014) implements the mean-mean, mean-sigma, Stocking-Lord, and Haebara methods for the transformation of IRT scales. Other functions in this package such as `est()` and `eap()` are used to estimate both item and ability parameter estimates. The **sirt** package (Robitzsch 2016) has the functions `equating.rasch()`, `equating.rasch.jackknife()`, `linking.haberman()`, and `linking.robust()`, all of which implement variants of IRT linking methods and many other functions for general IRT analysis. The packages **TAM** (Kiefer et al. 2016) and **eRm** (Mair and Hatzinger 2007) implement functions to estimate a variety of unidimensional and multidimensional IRT models.

We acknowledge the existence of the **plink** package (Weeks 2010), which unfortunately is no longer available on CRAN. The package is designed for linking mixed-format tests using IRT-based methods, and it also implements IRT true-score and observed-score equating. Interested readers can still download archived versions of the package at https://cran.r-project.org/src/contrib/Archive/plink/.

5.6 Summary

In this chapter, different methods of IRT equating have been described for the case when either IRT scores or sum scores are used for reporting the equating results. The methods have been illustrated using the **R** packages **SNSequate** and **equateIRT**.

The equating methods described require item parameter estimates from an IRT model as inputs, and these can be obtained from any IRT software package. We have illustrated how IRT models can be fitted using the **R** packages **ltm** and **mirt**. A list of alternative **R** packages for IRT analyses as well as methods of linking and equating were described in Sect. 5.5. It is worth mention that, although in this chapter we have only described item parameter linking and equating for IRT models using binary scored items, the methods can also be extended to accommodate polytomous score data (see, e.g., Kolen and Brennan 2014, Sect. 6.10). A method that can use either binary or polytomous scored data with kernel equating will be described in Chap. 7 together with a brief description of two IRT models for polytomous scored data which is given in Sect. 7.3.1. Further examples of IRT equating and linking can be found on the book's webpage.

References

Andersson, B., Bränberg, K., & Wiberg, M. (2013). Performing the kernel method of test equating with the package kequate. *Journal of Statistical Software, 55*(6), 1–25.

Andersson, B., & Wiberg, M. (2016). Item response theory observed-score kernel equating. *Psychometrika*. doi: 10.1007/s11336–016–9528–7.

Baker, F., & Kim, S. (2004). *Item response theory: Parameter estimation techniques*. New York: Marcel Dekker.

Bates, D., Mächler, M., Bolker, B., & Walker, S. (2015). Fitting linear mixed-effects models using lme4. *Journal of Statistical Software, 67*(1), 1–48.

Battauz, M. (2015). equateIRT: An R package for IRT test equating. *Journal of Statistical Software, 68*(7), 1–22.

Bechger, T., & Maris, G. (2015). A statistical test for differential item pair functioning. *Psychometrika, 80*(2), 317–340.

Birnbaum, A. (1968). Some latent trait models and their use in inferring any examinee's ability. In F. M. Lord & M. R. Novick (Eds.), *Statistical theories of mental test scores* (pp. 395–479). Reading: Adison-Wesley.

Chalmers, R. P. (2012). mirt: A multidimensional item response theory package for the R environment. *Journal of Statistical Software, 48*(6), 1–29.

Chen, M. (2004). Skewed link models for categorical response data. In M. Genton (Ed.), *Skew-elliptical distributions and their applications: A journey beyond normality* (Vol. 1, pp. 131–152). Boca Raton: Chapman & Hall/CRC.
Chen, M.-H., Dey, D. K., & Shao, Q.-M. (1999). A new skewed link model for dichotomous quantal response data. *Journal of the American Statistical Association, 94*(448), 1172–1186.
Cook, L. L., & Eignor, D. (1991). IRT equating methods. *Educational Measurement: Issues and Practice, 10*(3), 37–45.
De Boeck, P., Bakker, M., Zwitser, R., Nivard, M., Hofman, A., Tuerlinckx, F., & Partchev, I. (2011). The estimation of item response models with the lmer function from the lme4 package in R. *Journal of Statistical Software, 39*(12), 1–28.
De Boeck, P., & Wilson, M. (2004). *Explanatory item response models: A generalized linear and nonlinear approach.* New York: Springer.
DeMars, C. (2002). Incomplete data and item parameter estimates under JMLE and MML estimation. *Applied Measurement in Education, 15*(1), 15–31.
Estay, G. (2012). Characteristic curves scale transformation methods using asymmetric ICCs for IRT equating. Unpublished master's thesis, Department of Statistics, Pontificia Universidad Catolica de Chile.
Fischer, G., & Molenaar, I. (1995). *Rasch models: Foundations and recent developments.* New York: Springer.
González, J. (2014). SNSequate: Standard and nonstandard statistical models and methods for test equating. *Journal of Statistical Software, 59*(7), 1–30.
González, J., Wiberg, M., & von Davier A. A. (2016). A note on the Poisson's binomial distribution in item response theory. *Applied Psychological Measurement, 40*(4), 302–310.
Haebara, T. (1980). Equating logistic ability scales by a weighted least squares method. *Japanese Psychological Research, 22*, 144–149.
Hambleton, R. K., & Swaminathan, H. (1985). *Item response theory: Principles and applications.* Dordrecht: Kluwer Nijhoff Publishing.
Kiefer, T., Robitzsch, A., & Wu, M. (2016). *TAM: Test analysis modules.* R Package Version 1.995-0.
Kim, S. (2006). A comparative study of IRT fixed parameter calibration methods. *Journal of Educational Measurement, 43*(4), 355–381.
Kolen, M., & Brennan, R. (2014). *Test equating, scaling, and linking: Methods and practices* (3rd ed.). New York: Springer.
Lord, F. (1980). *Applications of item response theory to practical testing problems.* Hillsdale: Lawrence Erlbaum Associates.
Lord, F., & Novick, M. (1968). *Statistical theories of mental test scores.* Reading: Addison-Wesley.
Lord, F., & Wingersky, M. (1984). Comparison of IRT true-score and equipercentile observed-score "equatings". *Applied Psychological Measurement, 8*(4), 453–461.
Loyd, B. H., & Hoover, H. (1980). Vertical equating using the rasch model. *Journal of Educational Measurement, 17*(3), 179–193.
Mair, P., & Hatzinger, R. (2007). Extended Rasch modeling: The eRm package for the application of IRT models in R. *Journal of Statistical Software, 20*, 1–20.
Marco, G. L. (1977). Item characteristic curve solutions to three intractable testing problems. *Journal of Educational Measurement, 14*(2), 139–160.
Mislevy, R. J., & Bock, R. D. (1990). *BILOG 3: Item analysis and test scoring with binary logistic models.* Mooresville: Scientific Software International.
Ogasawara, H. (2000). Asymptotic standard errors of IRT equating coefficients using moments. *Economic Review (Otaru University of Commerce), 51*(1), 1–23.
Partchev, I. (2014). *Irtoys: Simple interface to the estimation and plotting of IRT models.* R Package Version 0.1.7.
Rizopoulos, D. (2006). ltm: An R package for latent variable modeling and item response theory analyses. *Journal of Statistical Software, 17*(5), 1–25.

Robitzsch, A. (2016). *sirt: Supplementary item response theory models*. R Package Version 1.12.2.

San Martín, E., González, J., & Tuerlinckx, F. (2015). On the unidentifiability of the fixed-effects 3PL model. *Psychometrika, 80*(2), 450–467.

Skaggs, G., & Lissitz, R. (1986). An exploration of the robustness of four test equating models. *Applied Psychological Measurement, 10*(3), 303.

Stocking, M., & Lord, F. (1983). Developing a common metric in item response theory. *Applied Psychological Measurement, 7*(2), 201–210.

Tuerlinckx, F., Rijmen, F., Molenberghs, G., Verbeke, G., Briggs, D., van den Noortgate, W., Meulders, M., & De Boeck, P. (2004). Estimation and software. In P. D. Boeck & M. Wilson (Eds.), *Explanatory item response models: A generalized linear and nonlinear approach* (Vol. 1, pp. 343–373). New York: Springer.

van der Linden, W. J. (Ed.) (2016). *Handbook of item response theory. Three volume set*. Boca Raton: Chapman and Hall/CRC.

van der Linden, W. J., & Barrett, M. (2016). Linking item response model parameters. *Psychometrika, 81*(3), 650–673.

von Davier, M., & von Davier, A. (2011). A general model for irt scale linking and scale transformations. In A. von Davier (Ed.), *Statistical models for test equating, scaling, and linking* (Vol. 1, pp. 225–242). New York: Springer.

Weeks, J. P. (2010). plink: An R package for linking mixed-format tests using IRT-based methods. *Journal of Statistical Software, 35*(12), 1–33.

Wiberg, M., van der Linden, W. J., & von Davier, A. A. (2014). Local observed-score kernel equating. *Journal of Educational Measurement, 51*, 57–74.

Wingersky, M. S., & Lord, F. M. (1984). An investigation of methods for reducing sampling error in certain IRT procedures. *Applied Psychological Measurement, 8*(3), 347–364.

Chapter 6
Local Equating

Abstract Local equating (van der Linden (2011) Local observed-score equating. In: von Davier A (ed) Statistical models for test equating, scaling, and linking. Springer, New York, pp 201–223) can be seen as an attempt to obtain a fairer equating in comparison to the traditional equating methods described in previous chapters. In this chapter, the concept of local equating is presented, and some existing local equating methods that are currently implemented in **kequate** (Andersson et al., J Stat Softw 55(6):1–25, 2013) and **SNSequate** (González, J Stat Softw 59(7):1–30, 2014) are illustrated.

6.1 The Concept of Local Equating

We motivate the idea behind local equating using the following hypothetical example[1] inspired by van der Linden and Wiberg (2010). Let $x_1, \ldots, x_{n_x}$ and $y_1, \ldots, y_{n_y}$ be observed score data for n_x and n_y test takers that have been administered test forms X and Y, respectively. Consider two test takers p and q with abilities θ_p and θ_q and with scores x_p and x_q in test form X, respectively. As we have seen in previous chapters, equating transformations are built from the corresponding score distributions $F_X(x)$ and $F_Y(y)$ (see Eq. (1.2)), which are estimated using the score data. Thus, to obtain an equivalent test score for x_p and x_q it is sufficient to evaluate $\varphi(x) = F_Y^{-1}(F_X(x))$ at these values to obtain $y_p = \varphi(x_p)$ and $y_q = \varphi(x_q)$, respectively. Suppose now that both p and q get the same test score so that $x_p = x_q$. Using the above approach, both test takers would obtain the same equated score. But what if test taker p is in reality more able than q? Would it be fair to assign him or her the same equated score as q? In general, would it be acceptable for test takers to obtain a score that actually depends on the abilities of other tests takers?

If the full score distributions for θ_p were available for both tests, $F_X(x|\theta_p)$ and $F_Y(y|\theta_p)$ and likewise for θ_q, $F_X(x|\theta_q)$ and $F_Y(y|\theta_q)$, a *fairer* equating would provide an equated value of x_p using $\varphi_p(x, \theta_p) = F_{Y|\theta_p}^{-1}(F_{X|\theta_p}(x))$. In the same vein, a fairer equated value for x_q can be obtained by building the equating transformation using

[1]Alternative motivations for local equating are given in Sect. B.6.

J. González, M. Wiberg, *Applying Test Equating Methods*,
Methodology of Educational Measurement and Assessment,
DOI 10.1007/978-3-319-51824-4_6

$F_X(x|\theta_q)$ and $F_Y(y|\theta_q)$. Thus, rather than a unique equating transformation, a family of equating transformations would be needed to assign fairer equated scores. This is the fundamental idea behind local equating.

6.1.1 True Equating Transformation

The ideas introduced above are well summarized by Lord (1980) who conceptualized the issue of fairness in equating by establishing the *equity* principle. Equity is fulfilled if the (conditional) probability function of scores on Y, $f_{y|\theta}$, and that of the transformed scores, $f_{\varphi(x)|\theta}$, are the same for every test taker at every ability level θ, i.e.,

$$f_{y|\theta} \equiv f_{\varphi(x)|\theta}. \tag{6.1}$$

Starting from the equity principle, and replacing probability functions with distribution functions, van der Linden (2013) derived what is called the family of *true equating transformations*

$$\varphi^*(x,\theta) = F_{Y|\theta}^{-1}(F_{X|\theta}(x)), \theta \in \Re, \tag{6.2}$$

where θ indexes the individual members of the family and $\Re$ is the set of possible values for θ (van der Linden 2000, 2013). The local equating method is thus different from the equating methods described in Chaps. 2, 3, 4, and 5 in that the latter utilize only *one* single equating transformation. Another immediate difference is that the equating transformation in Eq. (6.2) is built from the conditional distributions of the observed scores given θ.

Because the parameter indexing the family of equating transformations in Eq. (6.2) is not known, either an estimate or a proxy of this parameter must be used in practice to obtain an estimated equating transformation. For instance, when θ represents the ability of test takers under a response model, estimates of θ can be plugged into Eq. (6.2) to obtain an estimated equating transformation. It is thus possible to use, for example, maximum likelihood estimates or Bayesian expected a posterior estimates (van der Linden 2011). It is also possible to use an indirect estimate of θ that is some kind of *proxy* for the ability (van der Linden and Wiberg 2010). In this context, a proxy is defined as any monotone function of the ability that is measured (Wiberg and van der Linden 2011). For instance, under a NEAT design, information from the anchor test scores constitutes a suitable proxy of θ (van der Linden and Wiberg 2010). No matter the chosen type of value that is used for indexing the family of equating transformations, the aim of local equating is to obtain a transformation that is as close as possible to the true equating transformation.

van der Linden (2000) has shown that the family of true equating transformations fulfills all of the equating requirements defined in Chap. 1 (see Sect. 1.2.6). Several empirical studies have used the definition of the true equating transformation (van der Linden 2006a,b, 2011; Wiberg et al. 2014; Wiberg and van der Linden 2011; van der Linden and Wiberg 2010).

6.2 Performing Local Equating

A general algorithm to perform local equating was suggested by van der Linden (2006a), and this is summarized in the following three steps:

1. Use either an estimate or proxy of θ that is obtained from the test takers' response vectors.
2. Pick a true equating transformation at this estimate or proxy from the family in Eq. (6.2).
3. Use the chosen true transformation to calculate the equated score associated with the test taker's observed number-correct score.

One way to evaluate a performed local equating is to calculate the bias. The bias is defined as the difference between the actual obtained equating transformation $\varphi(x)$ and the true equating transformation $\varphi^*(x)$

$$\text{bias} = \varphi(x) - \varphi_x^*(x;\theta),\ \theta \in \Re \tag{6.3}$$

(van der Linden 2000; van der Linden and Wiberg 2010; Wiberg and van der Linden 2011). Another common measure that is used is the root mean squared error (RMSE).

6.3 Local Linear Equating Transformations

As it was in the case of traditional methods of equating, both equipercentile and linear transformations can be used under the local equating framework. The mathematical form of the equating transformation will depend on the equating design and sometimes on the type of index chosen for the family of equating transformations. In this section, we describe some local linear equating methods, detailing the type of index used and the most suitable equating design that the transformation is used for.

6.3.1 Local Linear Equating Conditioning on Anchor Test Scores: NEAT Design

Local linear equating conditioning on anchor test scores (Wiberg and van der Linden 2011) was the first proposed local linear method. Under a NEAT design, a suitable proxy of θ is the anchor test score, thus it is possible to condition on the anchor score and adjust the traditional linear equating transformation (see Eq. (3.2)) as follows

$$\varphi(x) = \mu_{Y|a} + \frac{\sigma_{Y|a}}{\sigma_{X|a}}(x - \mu_{X|a}),\ a \in \mathscr{A}. \tag{6.4}$$

where a is a possible value of the anchor test score and $\mathscr{A}$ is the entire set of possible values of anchor test scores. Estimates of $\mu_{Y|a}$, $\mu_{X|a}$, $\sigma_{X|a}$, and $\sigma_{Y|a}$ can be obtained in a straightforward manner from the test score data set.

6.3.2 Local Linear Equating Method of Conditional Means: SG Design

For the SG design, it is possible to use the observed score of the test form that is to be equated as a suitable proxy of θ. Thus, conditioning on the observed test score $X = x$ we have

$$\varphi(x) = \mu_{Y|x} + \frac{\sigma_{Y|x}}{\sigma_{X|x}}(x - \mu_{X|x}),\ x \in \mathscr{X}. \tag{6.5}$$

From classical test theory, using the fact that $\mu_{X|x} = x$ it is possible to reduce Eq. (6.5) to the local linear method of conditional means

$$\varphi(x) = \mu_{Y|x},\ x \in \mathscr{X}. \tag{6.6}$$

For a thorough discussion and empirical study of this method and local linear equating conditioning on anchor test scores, please refer to Wiberg and van der Linden (2011).

6.3.3 Local Linear Equating Examples in R

None of the local equating methods described above are currently part of any **R** package. However, in what follows we show that they can be easily implemented in **R**.

6.3.3.1 Implementing Local Linear Equating Conditioning on Anchor Test Scores

To illustrate local linear equating conditioning on anchor test scores, we use the KB data. Remember that the object KBneat is a list with two elements, x and y, containing the test score pairs (x, a) and (y, a) for test forms X and Y, respectively. To obtain conditional means and variances, it is thus necessary to first filter the data set for selected values of anchor scores. In this example we chose the values $a = 3, 6, 9, 12$.

```
> library(equate)

> dat.ax3<-KBneat$x[KBneat$x[,2]==3,]
> dat.ax6<-KBneat$x[KBneat$x[,2]==6,]
> dat.ax9<-KBneat$x[KBneat$x[,2]==9,]
> dat.ax12<-KBneat$x[KBneat$x[,2]==12,]

> dat.ay3<-KBneat$y[KBneat$y[,2]==3,]
> dat.ay6<-KBneat$y[KBneat$y[,2]==6,]
> dat.ay9<-KBneat$y[KBneat$y[,2]==9,]
> dat.ay12<-KBneat$y[KBneat$y[,2]==12,]
```

The resulting objects are two-column matrices containing the observed score values (first column) for a given anchor score (second column). For example, the observed realizations of the distribution $f_{X|A}(x|A = 12)$ looks like

```
> dat.ax12
     total anchor
6       36     12
129     30     12
221     34     12
799     33     12
1106    34     12
1359    34     12
1515    32     12
1569    33     12
```

The conditional means and standard deviations are calculated directly from the scores that are observed for a given anchor score. The following code can be used for this task:

```
> mu.x3<-mean(dat.ax3$total)
> mu.x6<-mean(dat.ax6$total)
> mu.x9<-mean(dat.ax9$total)
> mu.x12<-mean(dat.ax12$total)

> sigma.x3<-sd(dat.ax3$total)
> sigma.x6<-sd(dat.ax6$total)
> sigma.x9<-sd(dat.ax9$total)
> sigma.x12<-sd(dat.ax12$total)
```

```
> mu.y3<-mean(dat.ay3$total)
> mu.y6<-mean(dat.ay6$total)
> mu.y9<-mean(dat.ay9$total)
> mu.y12<-mean(dat.ay12$total)

> sigma.y3<-sd(dat.ay3$total)
> sigma.y6<-sd(dat.ay6$total)
> sigma.y9<-sd(dat.ay9$total)
> sigma.y12<-sd(dat.ay12$total)
```

Having the estimated conditional means and standard deviations, the equating transformation in Eq. (6.4) can be easily implemented. The following is an example of code that can be used to create a function called `local.lin()`, which takes as arguments the score values in $\mathscr{X}$ that are to be equated, and the estimated parameters $\mu_{Y|a}$, $\mu_{X|a}$, $\sigma_{X|a}$, and $\sigma_{Y|a}$. The function then returns the corresponding equated values in $\mathscr{Y}$.

```
> local.lin<-function(x,mu.y,mu.x,sigma.y,sigma.x){
+ phi.x<- mu.y+(sigma.y/sigma.x)*(x-mu.x)
+ return(phi.x)
> }
```

For example, we can create a table showing the equated values for the whole score scale for the case when $a = 3$ using the following code:

```
> eq.ll<-local.lin(0:36,mu.y3,mu.x3,sigma.y3,sigma.x3)
> tab.eqll<-cbind(0:36,eq.ll)
> tab.eqll
             eq.ll
 [1,]  0  0.1234318
 [2,]  1  1.2338482
 [3,]  2  2.3442647
 [4,]  3  3.4546811
 [5,]  4  4.5650975
 [6,]  5  5.6755140
 [7,]  6  6.7859304
 [8,]  7  7.8963468
 [9,]  8  9.0067632
[10,]  9 10.1171797
[11,] 10 11.2275961
[12,] 11 12.3380125
[13,] 12 13.4484290
[14,] 13 14.5588454
[15,] 14 15.6692618
[16,] 15 16.7796783
[17,] 16 17.8900947
[18,] 17 19.0005111
[19,] 18 20.1109275
[20,] 19 21.2213440
[21,] 20 22.3317604
[22,] 21 23.4421768
[23,] 22 24.5525933
[24,] 23 25.6630097
```

```
[25,] 24 26.7734261
[26,] 25 27.8838426
[27,] 26 28.9942590
[28,] 27 30.1046754
[29,] 28 31.2150918
[30,] 29 32.3255083
[31,] 30 33.4359247
[32,] 31 34.5463411
[33,] 32 35.6567576
[34,] 33 36.7671740
[35,] 34 37.8775904
[36,] 35 38.9880069
[37,] 36 40.0984233
```

Figure 6.1 shows a plot of the four equating transformations, and it was produced with the following code:

```
> plot(0:36,local.lin(0:36,mu.y3,mu.x3,sigma.y3,sigma.x3),
+ type='l', lty=1, lwd =2,col=1,ylab = expression(phi(x)),
+ xlab = "x")
> lines(0:36,local.lin(0:36,mu.y6,mu.x6,sigma.y6,sigma.x6),
+ lty = 2, lwd =2,col=2)
> lines(0:36,local.lin(0:36,mu.y9,mu.x9,sigma.y9,sigma.x9),
+ lty = 3, lwd =2,col=3)
> lines(0:36,local.lin(0:36,mu.y12,mu.x12,sigma.y12,sigma.x12),
+ lty = 4, lwd =2,col=4)
> legend("topleft",lty=c(1,2,3,4), col=c(1,2,3,4),
+ c("a=3","a=6","a=9","a=12"),lwd=c(2,2,2,2))
```

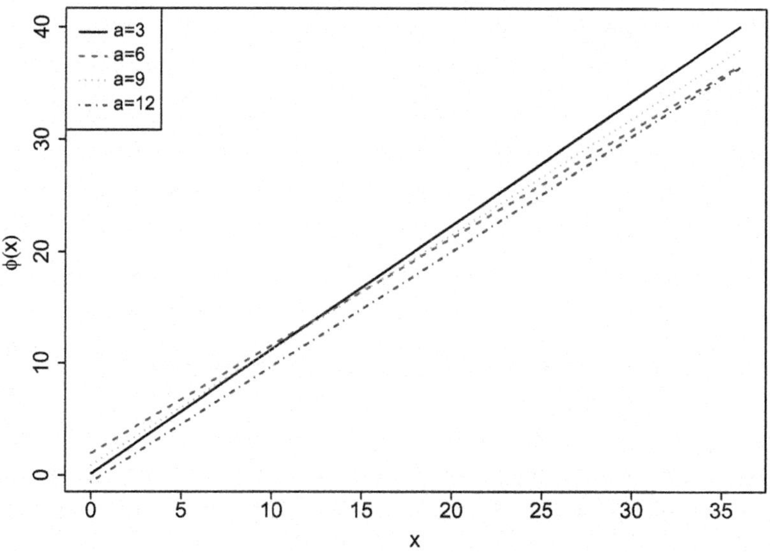

Fig. 6.1 True equating transformations for $a = (3, 6, 9, 12)$ for local linear equating

6.3.3.2 Implementing the Local Linear Equating Method of Conditional Means

An example of the local linear equating method of conditional means under the SG design is given next using the `ADM` data. Part of the **R** code described in Chap. 2 to create the objects `sgADM.x` and `sgADM.y` is used here to create the object `dat.xy`, which contains a two-column matrix with the score pairs (x, y) for each test taker.

```
> load(url("http://www.mat.uc.cl/~jorge.gonzalez/
+ EquatingRbook/ADM12.Rda"))
> sgADM.x <- apply(ADM12[,1:80],1,sum)
> sgADM.y <- apply(ADM12[,161:240],1,sum)
> dat.xy<-cbind(sgADM.x,sgADM.y)
```

As in the previous example, we first need to filter the data matrix. In the case of the SG design, we need to obtain the observed y scores for each of the observed values of x scores so that $\mu_{Y|x}$ can be calculated.

```
> dat.Allxy<-list()
> mu.xy<-c()
> for(i in 0:80){
>  dat.Allxy[[i+1]]<-dat.xy[dat.xy[,1]==i,]
>  mu.xy[i+1]<-mean(dat.Allxy[[i+1]][2])
> }
```

The object `mu.xy` contains the estimated values $\mu_{Y|x}$ with which an estimate of the equating transformation is obtained. Figure 6.2 shows the family of equating transformations obtained under the local linear equating method of conditional means. Note that for each value $X = x$, the equating transformation is actually a constant number corresponding to the conditional mean $\mu_{Y|x}$. Because the values $X = 0, 1, 2, 3, 4, 5, 6, 8, 9, 80$ were not observed, no equating transformation is plotted for these cases. The figure was produced using the following code.

```
> plot(0:80,mu.xy,type='p',ylab = expression(phi(x)),
+ xlab='x',lwd=2)
```

6.4 Local Equipercentile Equating Transformations

Some local equating methods that use an equipercentile-like transformation have been proposed in the literature. As was the case for local linear methods, their usability will depend on the type of index used to characterize the family of equating transformations, which is selected considering the equating data collection design. In this section we describe the methods that are currently implemented in the **R** packages **SNSequate** and **kequate**.

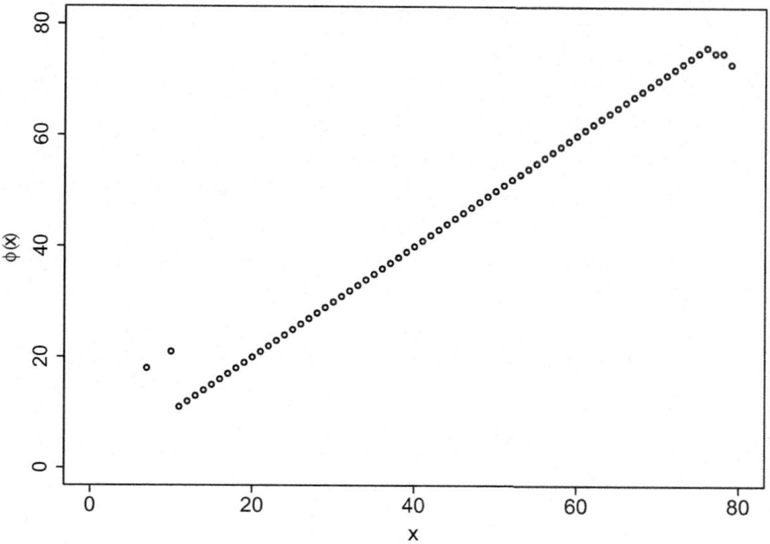

Fig. 6.2 Equating transformations for the local linear equating method of conditional means

6.4.1 Local IRT Observed-Score Equating

In local IRT observed-score equating[2] (van der Linden 2000, 2006a), the family of true equating transformations is indexed by the ability parameter underlying some IRT model. The two-parameter logistic (2PL) or the three-parameter logistic (3PL) IRT models have been shown to work well for local equating, in contrast to the one-parameter logistic (1PL) IRT model (von Davier et al. 2013). Both maximum likelihood and Bayesian expected a posterior estimates can be used to obtain $\hat{\theta}$ (van der Linden 2011). Another possibility is to use the full posterior distribution of θ for the test taker's response vector to obtain the posterior expectation of the true equating family (van der Linden 2000, 2006a). The family of local IRT observed-score equating transformations is defined as

$$\varphi(x) = F^{-1}_{Y|\hat{\theta}}(F_{X|\hat{\theta}}(x)), \theta \in \Re, \tag{6.7}$$

where the conditional distributions can be obtained using the recursive algorithm described in Lord and Wingersky (1984) (see Sect. B.5) or using alternative approaches as described in González et al. (2016). For more details about this method, we refer to van der Linden (2000, 2006a).

[2]This method is also called estimated conditional equating or estimated true equating.

6.4.2 Local Observed-Score Kernel Equating Conditioning on Anchor Test Scores

In Chap. 4, the observed-score kernel equating framework was described. In Wiberg et al. (2014), the local kernel equating conditional on anchor test score method was proposed for the NEAT design. This method has its roots in the local equating on anchor test score method described in van der Linden and Wiberg (2010). It relies on the assumption that the observed score on the common anchor test form A is an accurate proxy of the proficiency measured by test form X and test form Y. A discussion about the requirements that the anchor test should fulfill in order to be used as a proper proxy can be found in Wiberg and van der Linden (2011).

The family of local observed-score kernel equating transformations can be defined as

$$\varphi(x) = \hat{F}_{Y|a}^{-1}(\hat{F}_{X|a}(x)), \ a \in \mathscr{A}, \tag{6.8}$$

where $\hat{F}$ is the presmoothed and continuized conditional distribution obtained using the kernel equating framework (see Chap. 4).

6.4.3 Local IRT Observed-Score Kernel Equating

Another method is local IRT observed-score kernel equating (Wiberg et al. 2014; Andersson and Wiberg 2016), which relies on the assumption that the tests X and Y jointly fit an IRT model. The method has its origin in IRT observed-score equating and IRT observed-score kernel equating. From the scores in test forms X and Y, the distribution of number-correct scores given the ability θ can be obtained using either the Lord and Wingersky (1984) algorithm (see Sect. B.5) or alternative methods (González et al. 2016). After these distributions have been continuized, the family of equating transformations can be derived from them as

$$\varphi(x) = \tilde{F}_{Y|\hat{\theta}}^{-1}(\tilde{F}_{X|\hat{\theta}}(x)), \theta \in \Re, \tag{6.9}$$

where $\tilde{F}$ is the continuized distribution obtained using the general kernel equating framework described in Chap. 4. Note that because we are using equivalent test takers with the same level of ability, the design function used to obtain the estimated score probabilities is the identity function (see Eq. (B.1)).

6.4.4 *Local Equipercentile Equating Examples in R*

In the following sections, we give examples of local equipercentile equating methods that are currently implemented in the **R** packages **SNSequate** and **kequate**.

6.4.4.1 Local IRT Observed-Score Equating Using SNSequate

The function `le.eq()` in **SNSequate** can be used to perform local IRT observed-score equating. This function first estimates the conditional score distributions and then performs equipercentile equating using the estimated conditional distributions. A typical call reads as

```
le.eq(S.X, It.X, It.Y, Theta)
```

where the argument `S.X` is a vector containing the observed scores of the sample taking test form X. The arguments `It.X` and `It.Y` contain matrices of item parameter estimates coming from an IRT model fitted to test forms X and Y, respectively. These values are used together with the Lord-Wingersky recursive algorithm to obtain the conditional score distributions. The argument `Theta` can either be a number or vector of values representing the value of θ on which to condition on. The function returns as output a list containing the observed scores for each value of θ and the corresponding equated values.

The following example uses the parameter estimates in the `KB36` data to simulate item response data under a 3PL IRT model. We simulate two test forms that are both administered to a total of 2,500 individuals. From the two simulated test forms, sum scores are obtained and used as inputs in the `le.eq()` function to obtain the true equating transformation. The following code is used to store the values of the item parameters estimates and to generate the values of ability with which the response data will be generated.

```
> data("KB36", package = "SNSequate")
> Itx <- KB36$KBformX_par
> Ity <- KB36$KBformY_par
> Itx.m <- t(cbind(Itx[, 2], Itx[, 1], Itx[, 3]))
> Ity.m <- t(cbind(Ity[, 2], Ity[, 1], Ity[, 3]))
> Th <- rep(seq(-2, 2, 1), each = 500)
```

For the given values of item and person parameters stored in the objects `Itx.m`, `Itx.m`, and `Th`, respectively, the given values of probabilities used to produce Bernoulli random variables according to the 3PL probabilities (see Sect. 5.1) are obtained using the function `Pr()`. This function is used as an input in the function `Pattern()` to produce a full response pattern. These functions are created as follows.

```
> Pr <- function(theta, b, a = 1, c = 0)
+   c + (1-c)/(1+exp(-a*(theta-b)))
```

```
> Pattern <- function(theta, b, a = rep(1, length(b)),
+ c = rep(0, length(b)))
+ rbinom(length(b),1,Pr(theta, b, a, c))
```

For each value of $\theta = (-2.0, -1.0, 0.0, 1.0, 2.0)$, 500 response patterns are generated using the `DataGen()` function described below.

```
> DataGen <- function(Theta, ItemPar){
+   res <- NULL
+   N <- length(Theta)
+   for (i in 1 : N) res <- rbind(res, Pattern(Theta[i],
+   ItemPar[1, ], ItemPar[2, ], ItemPar[3, ]))
+   return(res)}
```

The final step to obtain the simulated data is to generate the 0/1 matrices of item responses that are stored in the objects `X` and `Y` for test forms X and Y, respectively. These matrices are then summed over columns to obtain the sum scores that are stored in the objects `sx` and `sy`.

```
> set.seed(10)
> X <- DataGen(Th, Itx.m)
> Y <- DataGen(Th, Ity.m)
> sx <- apply(X, 1, sum)
> sy <- apply(Y, 1, sum)
```

The actual local equating procedure is performed using the `le.eq()` function. To illustrate what the output looks like, suppose we want to obtain the results for test takers that have scored $X = 18$. The following code can be used:

```
> library(SNSequate)
> true <- le.eq(sx, Itx.m, Ity.m, Th)
> res  <- data.frame(Theta=true$Theta,Score=true$Obs.Sc,
+ Equated_value=true$resu)
> unique(res[res[, 2] == 18, ])
     Theta Score Equated_value
609     -1    18      20.13200
1001     0    18      20.67140
1509     1    18      20.66928
```

Because local equating generates a family of equating transformations, different test takers (i.e., with different values of θ) obtain different equated values, regardless of the fact that all of them have obtained the same sum score. This situation is exemplified by restricting the output to show only the results for test takers with $\theta = (-1, 0, 1)$, and where all of them obtained a sum score of 18.

A graphical display of the resulting family of equating transformation is shown in Fig. 6.3, where the equipercentile transformation has also been plotted for comparison. This figure was generated using the following code:

```
> resu <- vector("list", 9)
> ind <- seq(-2, 2, 1)
> for(i in 1 : 5){
+  resu[[i]] <- unique(res[res[,1] == ind[i],2:3])}
```

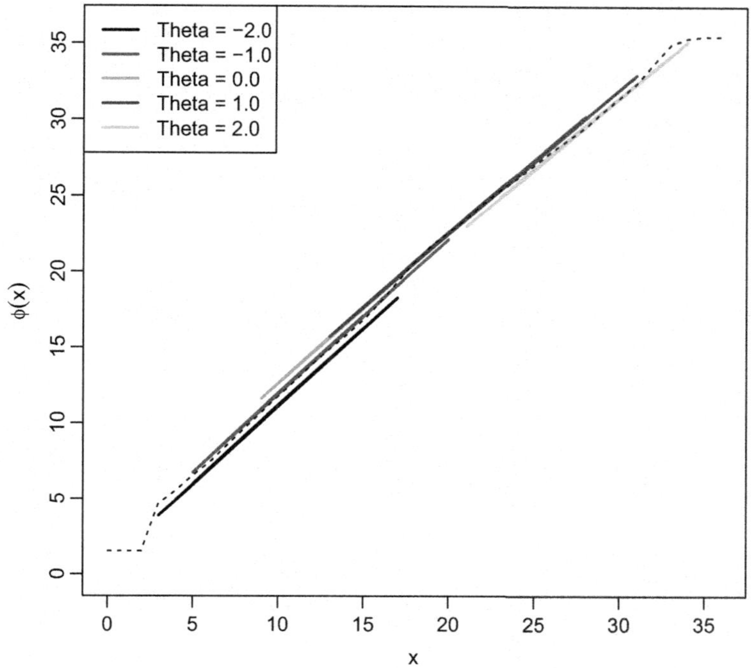

Fig. 6.3 True equating transformations for $\theta = (-2, -1, 0, 1, 2)$ using local IRT observed-score equating and equipercentile equating transformation (*dashed line*)

```
> est.eqp <- eqp.eq(sx, sy, 0 : 36)
> est.e <- est.eqp$resu[match(apply(X, 1, sum),
+ est.eqp$X)]
> plot(resu[[1]], type = 'l', xlim = c(0,36),
+ ylim = c(0, 36), col = 1, lwd = 2.0,
+ ylab = expression(phi(x)), xlab = "x")
> for(i in 2:5) lines(resu[[i]], type = 'l',
+ col=i, lwd = 2.0)
> lines(0 : 36, est.eqp$resu, type = 'l', lty = 2)
> legend("topleft", lty = c(1,1,1,1,1),
+ lwd = c(2,2,2,2,2),c("Theta = -2.0","Theta = -1.0",
+ "Theta = 0.0", "Theta = 1.0", "Theta = 2.0"),
+ col = c(1,2,3,4,5))
```

6.4.4.2 Local Observed-Score Kernel Equating Using kequate

To exemplify local kernel equating conditional on anchor test scores, we use the ADM test data described in Chap. 2 under a NEAT design. Similarly to the example given in Sect. 6.3.3.1, the score data need to be filtered for selected values of anchor scores. The following code shows how to perform this task for the case $a = 20$ and using the objects `neat.X` and `neat.Y` created in Sect. 2.2.4.

```
> dneat.X <- as.data.frame(neat.X)
> dneat.Y <- as.data.frame(neat.Y)
> Adm20P <- dneat.X$verb.x[dneat.X$verb.xa==20]
> Adm20Q <- dneat.Y$verb.y[dneat.Y$verb.ya==20]
```

Once the observed score values for a given anchor score are obtained (objects `Adm20P` and `Adm20Q`), the obtained data are analyzed using the five steps of kernel equating described in Chap. 4.[3]

```
> fXAdm20 <- kefreq(Adm20P,0:80)
> fYAdm20 <- kefreq(Adm20Q,0:80)
> glmXAdm20 <- glm(frequency~I(X)+I(X^2),data=fXAdm20,
+ family=poisson, x = TRUE)
> glmYAdm20 <- glm(frequency~I(X)+I(X^2),data=fYAdm20,
+ family = poisson, x = TRUE)
> lA20 <- kequate("EG",0:80,0:80,glmXAdm20,glmYAdm20)
```

Equated values can be obtained using the `getEq()` function and stored in an object called `A20` as follows:

```
> A20  <- getEq(lA20)
> A20
 [1] -2.36134259 -2.01951704 -1.70979658 -1.38579527
 [5] -1.02357764 -0.59797660 -0.06874029  0.62456646
 [9]  1.51687911  2.55058022  3.64847963  4.77118536
[13]  5.90316748  7.03863746  8.17544443  9.31277541
[17] 10.45031465 11.58793617 12.72558983 13.86325719
[21] 15.00093320 16.13861586 17.27630414 18.41399444
[25] 19.55168336 20.68936779 21.82704897 22.96472888
[29] 24.10241186 25.24009885 26.37778907 27.51547978
[33] 28.65316736 29.79085060 30.92853090 32.06621154
[37] 33.20389551 34.34158381 35.47927452 36.61696430
[41] 37.75465033 38.89233220 40.03001211 41.16769363
[45] 42.30537889 43.44306820 44.58075891 45.71844757
[49] 46.85613202 47.99381277 49.13149268 50.26917505
[53] 51.40686167 52.54455152 53.68224148 54.81992825
[57] 55.95761039 57.09528924 58.23296746 59.37064801
[61] 60.50833093 61.64601310 62.78368896 63.92135177
[65] 65.05899393 66.19660466 67.33416202 68.47161979
[69] 69.60888051 70.74574492 71.88180002 73.01617420
[73] 74.14698465 75.27008950 76.37632293 77.44575263
[77] 78.43807022 79.29044113 79.97019181 80.53221703
[81] 81.08812053
```

Repeating the above analysis with similar names for the objects for other anchor test score values, e.g., $a = 15$ and $a = 20$, a plot of the different obtained equating transformations can be produced using the code below. The resulting plot is shown in Fig. 6.4.

[3]Because we compare the results for test taker groups having the same anchor test score, test scores can be equated assuming an EG design.

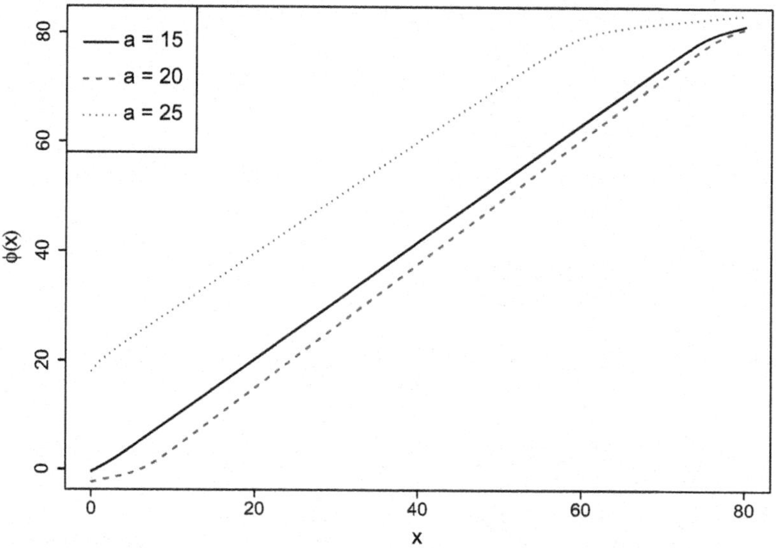

Fig. 6.4 Local observed-score kernel equating for anchor test scores 15, 20, and 25

```
> plot(0:80,A15, type='l',lwd=2,col=1,
+ ylab = expression(phi(x)),xlab = "x")
> lines(0:80,A20, lty=2,lwd=2,col=2)
> lines(0:80,A25,lty=3,lwd=3,col=3)
> legend("topleft", lty = c(1, 2, 3),
+ lwd = c(2, 2, 3),c("a = 15", "a = 20",
+ "a = 25"),col = c(1, 2, 3))
```

The anchor test used in the ADM data is external. Because equating under a NEAT design can be performed using either an internal or external anchor test, the equating transformation might have to be adjusted accordingly. Two examples of such adjustments for the case of internal anchor tests can be found in Wiberg and van der Linden (2011).

6.4.4.3 Local IRT Observed-Score Kernel Equating Using kequate

To illustrate local IRT observed-score kernel equating with **kequate** we use the `DataGen()` function described in the previous section to simulate data under a NEAT design. Two test forms (both containing 10 items) and an anchor test also containing 10 items are administered to two groups of 1,000 test takers each. The item response data are assumed to have been generated by a 2PL model. Given values for discrimination parameters are generated from uniform distributions whereas both item parameters and ability parameters are generated from normal distributions. The code used to perform these tasks is

```
> set.seed(5)
> aX <- runif(10,0.3,1.5)
> aY <- runif(10,0.3,1.5)
> aA <- runif(10,0.3,1.5)
> bX <- rnorm(10)
> bY <- rnorm(10)
> bA <- rnorm(10)
> thetaP <- rnorm(1000,mean=0.5)
> thetaQ <- rnorm(1000,mean=0)

> itx<-rbind(c(bX,bA),c(aX,aA),c(0,0))
> ity<-rbind(c(bY,bA),c(aY,aA),c(0,0))

> tP <- DataGen(thetaP, itx)
> tQ <- DataGen(thetaQ, ity)
```

where the objects `tP` and `tQ` contain the simulated 0/1 matrices of item responses.

As will be seen in Chap. 7, the function `irtose()` in **kequate** is used to perform IRT observed-score kernel equating. However, the same function with a small modification in one of the arguments also permits one to perform local IRT observed-score kernel equating. In this section we only describe the necessary arguments of `irtose()` needed to perform local IRT observed-score kernel equating, postponing more details of this function to Chap. 7. A typical call with the minimal arguments needed to perform local IRT observed-score kernel equating is as follows:

```
irtose(design, P, Q, x, y, a, qpoints,...)
```

The argument `design` is used to set the equating design under which the equating is to be performed. Current options are `"EG"`, `"PSE"` and `"CE"` for the EG design, and poststratification and chained equating methods under a NEAT design, respectively. The arguments `x`, `y`, and `a` are used to specify the score scales for test forms X, Y, and the anchor test form A, respectively. `P` and `Q` are most commonly IRT model objects[4] created with the package **ltm** from which item parameter estimates and other useful information is obtained and passed to `irtose()`. It is also possible to use the **mirt** package instead of the **ltm** package. The argument `qpoints` is actually used to integrate out the ability distributions as used in IRT observed-score equating (see Chap. 5). However, when set to a particular value, the result is a performed equating using the conditional score distributions at that fixed value.

As an example, suppose we want to know the equating transformation for test takers with $\theta = (-2, 0, 2)$. The following code shows first how the 2PL IRT model is fitted with the **ltm** package using the simulated item response data (objects `tP`

[4]In addition, 0/1 matrices of item responses can be used as input in which case IRT models are internally fitted.

and tQ) and then how these objects are passed to irtose() to perform local IRT observed-score kernel equating under the poststratification approach.[5]

```
> library(ltm)
> tPm <- ltm(tP~z1, IRT.param=FALSE)
> tQm <- ltm(tQ~z1, IRT.param=FALSE)

> library(kequate)
> Low <- irtose("PSE",tPm,tQm,0:10,0:10,0:10,
+ model="2pl",qpoints=-2)
> Mid <- irtose("PSE",tPm,tQm,0:10,0:10,0:10,
+ model="2pl",qpoints=0)
> High<- irtose("PSE",tPm,tQm,0:10,0:10,0:10,
+ model="2pl",qpoints=2)
```

The created objects Low, Mid, and High contain the performed local IRT observed-score kernel equating transformation picked at θ values of -2, 0, and 2, respectively. A plot of these equating transformations is shown in Fig. 6.5, which was produced using the code below.

```
> low  <- getEq(Low)
> mid  <- getEq(Mid)
> high <- getEq(High)
```

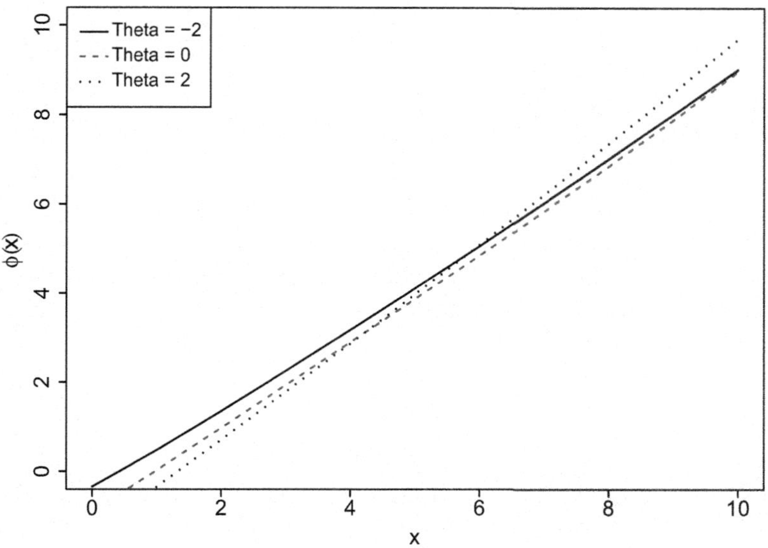

Fig. 6.5 Local IRT observed-score kernel equating for $\theta = -2$, 0 and 2

[5]An example showing local IRT observed-score kernel equating under the chained equating approach is shown in Andersson and Wiberg (2014).

```
> plot(0:10,low, type='l',lwd=2,col=1,ylab =
+ expression(phi(x)),xlab = "x",ylim=c(0,10))
> lines(0:10,mid, lty = 2,lwd=2,col=2)
> lines(0:10,high,lty = 3,lwd=3,col=1)
> legend("topleft", lty = c(1, 2, 3),lwd = c(2, 2, 3),
+ c("Theta = -2", "Theta = 0", "Theta = 2"),col = c(1, 2, 1))
```

6.5 Other Local Equating Methods

In the previous sections, we have described and exemplified only those local equating methods that are currently implemented in some **R** packages. However, several other local equating methods have been proposed in the literature. These methods include the posterior expected true score equating (van der Linden 2006a,b), local equating with ability estimated from anchor test scores (Janssen et al. 2009), local kernel equating with ability estimated from the anchor test (Wiberg et al. 2014), local observed-score equating with anchor test design using the full distribution (van der Linden and Wiberg 2010), and the two recently proposed local linear IRT observed-score equating and local linear kernel IRT observed score equating methods (Wiberg 2016). Because research on local equating is an active field and new methods are continuously emerging, we expect that more local equating methods will probably be included in different **R** packages in the near future.

6.6 Summary

This chapter discussed the general concept of local equating. Implementations of some local linear equating methods were illustrated using the **R** programming environment, and some local equipercentile equating methods were illustrated using the **R** packages **kequate** and **SNSequate**.

References

Andersson, B., Bränberg, K., & Wiberg, M. (2013). Performing the kernel method of test equating with the package kequate. *Journal of Statistical Software, 55*(6), 1–25.

Andersson, B., & Wiberg, M. (2014). *IRT observed-score kernel equating with the R package kequate*. R: Vignette. Retrieved from https://cran.r-project.org/web/packages/kequate/vignettes/irtguide.pdf

Andersson, B., & Wiberg, M. (2016). Item response theory observed-score kernel equating. *Psychometrika*. doi: 10.1007/s11336-016-9528-7.

González, J. (2014). SNSequate: Standard and nonstandard statistical models and methods for test equating. *Journal of Statistical Software, 59*(7), 1–30.

González, J., Wiberg, M., & von Davier, A. A. (2016). A note on the Poisson's binomial distribution in item response theory. *Applied Psychological Measurement, 40*(4), 302–310.
Janssen, R., Magis, D., San Martín, E., & Del Pino, G. (2009). Local equating in the NEAT design. Paper presented at the National Council on Measurement in Education, San Diego.
Lord, F. (1980). *Applications of item response theory to practical testing problems*. Hillsdale: Lawrence Erlbaum Associates.
Lord, F., & Wingersky, M. (1984). Comparison of IRT true-score and equipercentile observed-score "equatings". *Applied Psychological Measurement, 8*(4), 453–461.
van der Linden, W. J. (2000). A test-theoretic approach to observed-score equating. *Psychometrika, 65*(4), 437–456.
van der Linden, W. J. (2006a). Equating error in observed-score equating. *Applied Psychological Measurement, 30*(5), 355–378.
van der Linden, W. J. (2006b). Equating scores from adaptive to linear tests. *Applied Psychological Measurement, 30*(6), 493–508.
van der Linden, W. J. (2011). Local observed-score equating. In A. von Davier (Ed.), *Statistical models for test equating, scaling, and linking* (pp. 201–223). New York: Springer.
van der Linden, W. J. (2013). Some conceptual issues in observed-score equating. *Journal of Educational Measurement, 50*(3), 249–285.
van der Linden, W. J., & Wiberg, M. (2010). Local observed-score equating with anchor-test designs. *Applied Psychological Measurement, 34*(8), 620–640.
von Davier, M., González, J., & von Davier, A. A. (2013). Local equating using the Rasch model, the OPLM, and the 2PL IRT model -or- what is it anyway if the model captures everything there is to know about the test takers? *Journal of Educational Measurement, 50*(3), 295–303.
Wiberg, M. (2016). Alternative linear item response theory observed-score equating methods. *Applied Psychological Measurement, 40*(3), 180–199.
Wiberg, M., & van der Linden, W. J. (2011). Local linear observed-score equating. *Journal of Educational Measurement, 48*, 229–254.
Wiberg, M., van der Linden, W. J., & von Davier, A. A. (2014). Local observed-score kernel equating. *Journal of Educational Measurement, 51*, 57–74.

Chapter 7
Recent Developments in Equating

Abstract This chapter briefly describes some recent developments in test equating and provides examples of how they are performed using the **R** packages **kequate** (Andersson et al., J Stat Softw 55(6):1–25, 2013) and **SNSequate** (González, J Stat Softw 59(7):1–30, 2014). The chapter begins with recent developments within the kernel equating framework, including different bandwidth selection methods, the use of different kernels in the continuization step, and IRT kernel equating. A Bayesian approach to test equating and the assessment of equating transformations are also discussed. The chapter ends with some concluding reflections on the future of equating research connected to the use of **R**.

7.1 Alternative Kernel Equating Transformations

We have seen in Chap. 4 that both **kequate** and **SNSequate** implement the uniform kernel and the logistic kernel as alternatives to the Gaussian kernel (see Sects. 4.4.2, 4.4.3, and 4.4.4). Two other recently suggested alternative kernels for equating are the Epanechnikov kernel and the adaptive kernel (Cid and von Davier 2015; González and von Davier 2017), both of which are implemented in **SNSequate** and are briefly described next.

7.1.1 Epanechnikov Kernel

The continuization step (see Sect. 4.4) involves the use of a continuous random variable that characterizes the kernel to be used for equating. The Epanechnikov kernel (Epanechnikov 1969) is defined for a random variable V with the density function

$$f(v) = \frac{3}{4}(1 - v^2)1_{|v|\leq 1}, \tag{7.1}$$

J. González, M. Wiberg, *Applying Test Equating Methods*,
Methodology of Educational Measurement and Assessment,
DOI 10.1007/978-3-319-51824-4_7

and corresponding CDF

$$F(v) = \begin{cases} 0 & v < -1 \\ \dfrac{3v - v^3 + 2}{4} & -1 \leq v \leq 1 \\ 1 & v > 1 \end{cases},$$

for which it is easily verified that $E(V) = 0$ and $Var(V) = \frac{1}{5}$. Following the steps described in Sect. 4.4, we have that the Epanechnikov kernel continuization becomes

$$F_{hX} = \sum_{\mathscr{J}} \frac{r_j \left(3R_{jX}(x) - R_{jX}^3(x) + 2\right)}{4} + \sum_{\mathscr{K}} r_j, \tag{7.2}$$

where $R_{jX}(x) = \frac{x - a_X x_j - (1-a_X)\mu_X}{a_X h_X}$, $\mathscr{J}$ is the set of all j such that $-1 \leq R_{jX} \leq 1$, and $\mathscr{K}$ is the set of j such that $R_{jX} > 1$. Note that the penalty function method for the selection of the bandwidth parameter (see Eq. (4.11)) applies straightforwardly. The needed density estimate and the derivatives used in the second summand in the penalty function, are given in Sect. B.7.

7.1.2 Adaptive Kernels

Adaptive kernels (e.g., Silverman 1986) allow the bandwidth parameter h_X to vary across the data points in the score distribution. The kernel continuization has the form

$$F_{h_{jX}}(x) = \sum_j r_j K\left(\frac{x - a_{jX}x_j - (1 - a_{jX})\mu_x}{a_{jX}h_{jX}}\right),$$

where $a_{jX} = \frac{\sigma_X^2}{\sigma_X^2 + h_{jX}^2}$ and $h_{jX} = \lambda_j h_X$, $(j = 1, \ldots, J)$. For illustration, we will consider a Gaussian adaptive kernel so that $K(\cdot) = \Phi(\cdot)$.

Silverman (1986) suggested the following steps to obtain λ_j,

i. Find a pilot estimate of the density, $\widetilde{f}(t)$, such that $\widetilde{f}(X_j) > 0 \; \forall j$

ii. Define a local bandwidth factor λ_j as

$$\lambda_j = \left(\frac{\widetilde{f}(X_j)}{g}\right)^{-\alpha},$$

where g is the geometric mean of $\widetilde{f}(X_j)$ and α is a sensibility parameter satisfying $0 \leq \alpha \leq 1$. Silverman's recommendation is to use $\alpha = 0.5$.

To obtain λ_j we propose as a pilot estimate

$$\widetilde{f}(x) = \sum_j r_j \phi\left(\frac{x - a_X x_j - (1 - a_X)\mu_x}{a_X h_X}\right)\frac{1}{a_X h_X}, \tag{7.3}$$

where h_X can be obtained using any of the bandwidth selection methods described in Sects. 4.4.1, 7.2.1, and 7.2.2.

Following the strategy described above, we can also obtain $F_{h_{jY}}$, such that the adaptive kernel equating transformation becomes $\varphi(x) = F_{h_{jY}}^{-1}(F_{h_{jX}}(x))$.

7.1.3 *Examples of Epanechnikov and Adaptive Kernel Equating in SNSequate*

The `ker.eq()` function described in Sect. 4.5.1 implements both the Epanechnikov and Gaussian adaptive kernels by setting the arguments `kert="epan"` and `kert="adap"`, respectively. The sensitivity parameter α is specified using the argument `alpha`, and the value of the bandwidth parameter h_X for the pilot estimate in Eq. (7.3) is set using the argument `h.adap`. If no value is provided for these two arguments, then `alpha` is set to 0.5 and h_X will be automatically estimated using the penalty method described in Sect. 4.4.1.

The following example shows a comparison of equated values obtained from five different kernel equating transformations using the `Math20EG` data. The results for the five different kernel equating transformations are stored in the objects `mod.gauss`, `mod.logis`, `mod.unif`, `mod.epan`, and `mod.adap` for the Gaussian, logistic, uniform, Epanechnikov, and adaptive kernel continuizations, respectively.

```
> data("Math20EG")
> mod.gauss = ker.eq(Math20EG,degree=c(2,3),design="EG",
+ kert="gauss")
> mod.logis = ker.eq(Math20EG,degree=c(2,3),design="EG",
+ kert="logis")
> mod.unif  = ker.eq(Math20EG,degree=c(2,3),design="EG",
+ kert="unif")
> mod.epan  = ker.eq(Math20EG,degree=c(2,3),design="EG",
+ kert="epan")
> mod.adap  = ker.eq(Math20EG,degree=c(2,3),design="EG",
+ kert="adap")
```

A table showing the equated values for the five kernel equating transformations that are stored in the object `ker.comp` can be obtained using the following code:

```
> ker.comp<-cbind(Scale=0:20,Gauss=mod.gauss$eqYx,
+ Logistic=mod.logis$eqYx, Uniform=mod.unif$eqYx,
+ Epanechnikov=mod.epan$eqYx, Adaptive=mod.adap$eqYx)
> round(ker.comp,4)
   Scale   Gauss Logistic Uniform Epanechnikov Adaptive
 [1,] 0   0.3937   0.4474  0.4392       0.4197   0.5697
 [2,] 1   1.5813   1.5732  1.6387       1.5529   1.6070
 [3,] 2   2.6404   2.6285  2.6783       2.6158   2.6219
 [4,] 3   3.6444   3.6353  3.6762       3.6285   3.6165
 [5,] 4   4.6316   4.6253  4.6604       4.6224   4.6082
 [6,] 5   5.6178   5.6137  5.6434       5.6141   5.6094
 [7,] 6   6.6100   6.6079  6.6313       6.6111   6.6187
 [8,] 7   7.6120   7.6116  7.6280       7.6168   7.6342
 [9,] 8   8.6260   8.6269  8.6361       8.6331   8.6583
[10,] 9   9.6530   9.6549  9.6576       9.6608   9.6929
[11,] 10 10.6935  10.6960 10.6937      10.7007  10.7371
[12,] 11 11.7471  11.7497 11.7448      11.7527  11.7878
[13,] 12 12.8126  12.8147 12.8097      12.8151  12.8434
[14,] 13 13.8869  13.8879 13.8849      13.8840  13.9047
[15,] 14 14.9641  14.9633 14.9638      14.9534  14.9723
[16,] 15 16.0339  16.0305 16.0349      16.0128  16.0414
[17,] 16 17.0781  17.0717 17.0805      17.0443  17.0959
[18,] 17 18.0677  18.0578 18.0729      18.0197  18.1021
[19,] 18 18.9607  18.9520 18.9702      18.9418  19.0032
[20,] 19 19.7183  19.7339 19.7072      19.8442  19.7891
[21,] 20 20.3930  20.4613 20.2781      20.7254  20.4913
```

The obtained equated values are quite similar. To see how the different kernels differ in terms of the standard error of equating (SEE), Fig. 7.1 shows the SEE for each equated value using the five kernel equating transformations. Some differences are observed in terms of SEE, especially at the extremes of the score scale. This figure was created using the following code:

```
> score <- 0:20
> plot(score,mod.gauss$SEEYx,type="l",col=1,lwd=2,
+ xlab="Scores",ylab="SEE",ylim=c(0,0.4),main="")
> lines(score,mod.logis$SEEYx,col=2,lwd=2)
> lines(score,mod.unif$SEEYx,col=3,lwd=2)
> lines(score,mod.epan$SEEYx,col=4,lwd=2)
> lines(score,mod.adap$SEEYx,col=5,lwd=2)
> legend("topright", col = c(1,2,3,4,5),
+ c("Gaussian","Logistic","Uniform","Epanechnikov",
+ "Adapative"), lty=c(1,1,1,1,1),lwd=2)
```

We point out that when performing kernel equating with the Epanechnikov and adaptive kernels, other functions described in previous chapters such as `PREp()`, to obtain percent relative errors, and `SEED()`, to obtain standard error of equating differences, can also be used. Further examples showing the use of these functions are available on the book's webpage.

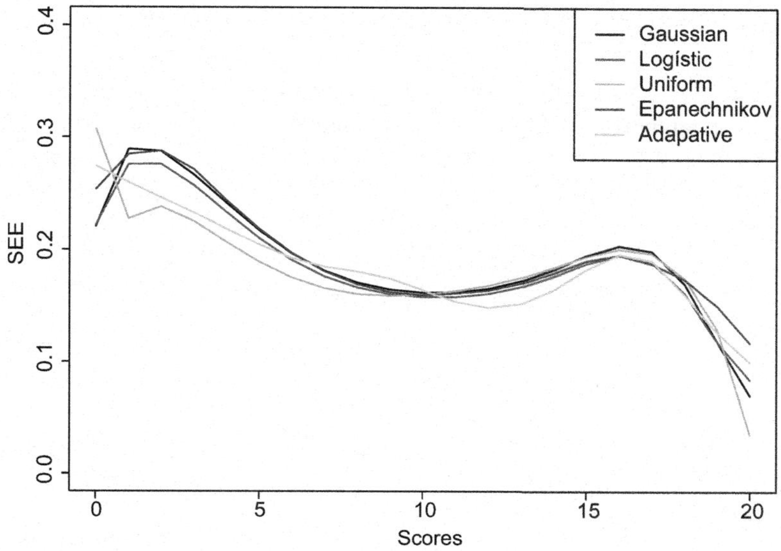

Fig. 7.1 SEE for five kernel equating transformations

7.2 Bandwidth Selection in Kernel Equating

The penalty function method described in Sect. 4.4 was the first approach suggested to obtain the bandwidth parameter in the kernel equating framework. Other alternative bandwidth selection methods in kernel equating have recently been proposed (Liang and von Davier 2014; Andersson and von Davier 2014; Häggström and Wiberg 2014). In what follows, we describe two of the newly proposed methods that are currently implemented in **kequate**.

7.2.1 Rule-Based Bandwidth Selection

Andersson and von Davier (2014) proposed an adjusted version of Silverman's rule of thumb (Silverman 1986), which states that for a standard Gaussian kernel, an appropriate value for the bandwidth parameter is $h_X = .9\sigma_X n_X^{-1/5}$. In the context of kernel equating, the modification of this general rule leads to selection of the bandwidth parameter as follows:

$$h_X = \frac{9\sigma_X}{\sqrt{100 n_X^{2/5} - 81}}. \tag{7.4}$$

7.2.2 Bandwidth Selection with Double Smoothing

Häggström and Wiberg (2014) proposed the use of double smoothing (DS, Hall et al. 1992) to select the bandwidth in kernel equating. The overall idea of DS is to select the bandwidth that minimizes an estimate of the targeted estimator's mean squared error where a pilot bandwidth is used to estimate the bias part. The goal of the final DS estimate is to preserve the characteristics of the estimated relative frequency distribution. The resulting objective function is similar to the first summand used in the penalty method described in Sect. 4.4.1 (see Eq. (4.11)) and reads as

$$\mathrm{DS}(h_X) = \sum_l (\hat{r}_l^* - \hat{f}_{h_X}^*(x_l^*))^2, \tag{7.5}$$

where $\hat{r}_l^*$ and $\hat{f}_{h_X}^*(x_l^*)$ are defined in Sect. B.8. In practice, the following steps are carried out[1]

1. Start with a large subjectively chosen bandwidth g_X and estimate f_{g_X} at the score values and the values halfway between the score values.
2. Because the smooth estimate obtained in the previous step will not perfectly interpolate the estimated score probabilities; the $\hat{r}$:s, this can be improved by estimating f_{h_X} at the actual score values, i.e. we can obtain a DS estimate $\hat{f}_{h_X}^*$.
3. Select the bandwidth that minimizes Eq. (7.5).

7.2.3 Examples of the Rule-Based and Double Smoothing Bandwidth Selection Methods Using kequate

The rule-based and the DS bandwidth selection methods are implemented in the `kequate()` function by setting the arguments `altopt=TRUE` and `DS=TRUE`, respectively. In the following example, we use the admissions data and the objects `egADMx` and `egADMy` created in Sect. 4.2.2.1.

```
> egADMrb <- kequate("EG",0:80,0:80,
+ egADMx, egADMy,altopt=TRUE)
> getH(egADMrb)
```

The `getH()` function is used to extract the value of the bandwidth parameter, which in this case yields $h_X = 1.832854$ and $h_Y = 1.765363$ for the X and Y score distributions, respectively.

In a very similar way, we can retrieve the bandwidth parameter obtained using the DS method, as illustrated in the following code:

[1] More details are given in Sect. B.8.

```
> egADMds <- kequate("EG",0:80,0:80,
+ egADMx, egADMy,DS=TRUE)
> getH(egADMds)
```

yielding the optimal bandwidths 0.6032495 for h_X and 0.5636444 for h_Y.

7.3 Item Response Theory Kernel Equating

We saw in Chap. 4 that score probabilities were calculated using the design functions after presmoothing the data using log-linear models. On the other hand, it was pointed out in Sect. 5.3.2 (see also Sect. B.5) that the (Lord and Wingersky 1984) algorithm can use the item parameter estimates from an IRT model to obtain conditional score distributions by calculating the corresponding score probabilities for each score on a test. This means that IRT models together with the Lord-Wingersky method can be used as an alternative to log-linear presmoothing and the use of design functions for the estimation of score probabilities. IRT kernel equating (Andersson and Wiberg 2014, 2016) is based on this alternative. More specifically, if we consider a Gaussian kernel and an IRT model $\pi(\theta, \omega_j)$ fitted to score data from test form X, then the estimated F_X is defined as

$$\hat{F}_{h_X}(x) = \sum_j \hat{r}_j(\theta)\Phi\left(\frac{x - a_X x_j - (1 - a_x)\mu_X}{a_X h_X}\right), \tag{7.6}$$

where $\hat{r}_j(\theta) = \widehat{\Pr(X = x_j} \mid \theta)$. In order to find the marginal score probabilities, the Lord-Wingersky algorithm and an extension of the algorithm (Thissen et al. 1995) are used together with a quadrature distribution for θ for binary and polytomous score data, respectively.

In Chap. 5, IRT models for binary scored items were discussed. In the following section we briefly describe two commonly used IRT models for the case when items are scored in more than two categories.

7.3.1 *Two Polytomous IRT Models*

The graded response model (Samejima 1969) is used to model items scored in ordered polytomous categories, for instance, items scored using a Likert scale. The probability of a randomly chosen test taker with ability θ scoring in category $x \in \{1, \ldots, m_j\}$ on item j is defined as

$$P_{j,x}(\theta) = \begin{cases} 1 - \dfrac{1}{1+\exp(-a_j(\theta - b_{j,x+1}))}, & x = 1 \\ \dfrac{1}{1+\exp(-a_j(\theta - b_{j,x}))} - \dfrac{1}{1+\exp(-a_j(\theta - b_{j,x+1}))}, & 2 \le x < m_j \ , \\ \dfrac{1}{1+\exp(-a_j(\theta - b_{j,x}))}, & x = m_j \end{cases} \quad (7.7)$$

where $b_{j,x}$ and a_j are threshold and item discrimination parameters, respectively.

The generalized partial credit model (Muraki 1992), on the other hand, accommodates partial credit data where the item responses are scored in $\{1, \ldots, S\}$ with S being the highest score category for the item. The probability that a randomly chosen test taker with ability θ scores x on item j is defined as

$$P_{j,x}(\theta) = \frac{\exp(\sum_{v=2}^{x} a_j(\theta - b_{j,v}))}{\sum_{c=2}^{m_j} \exp(\sum_{v=2}^{c} a_j(\theta - b_{j,v}))}. \quad (7.8)$$

where the $b_{j,v}$ terms are referred to as step difficulty parameters and $\sum_{v=1}^{S}(\theta - b_{j,v}) \equiv 0$.

7.3.2 *Performing IRT Kernel Equating with kequate*

The function `irtose()` was briefly introduced in Sect. 6.4.4.3 to perform local IRT observed-score kernel equating. In this section, we give a more detailed description of this function. Subsequent sections give examples of using `irtose()` to perform IRT observed-score kernel equating for both binary and polytomous score data.

The general function call for `irtose()` is

```
irtose(design="CE", P, Q, x, y, a = 0, qpoints=
seq(-6, 6, by=0.1),model="2pl", catsX=0, catsY=0,
catsA=0, see="analytical",replications = 50,
kernel="gaussian", h = list(hx=0, hy=0, hxP=0,
haP=0, hyQ=0, haQ=0), hlin = list(hxlin = 0,
hylin = 0, hxPlin = 0,haPlin = 0,hyQlin = 0,
haQlin = 0), KPEN=0, wpen=0.5, linear = FALSE,
slog = 1, bunif = 1, altopt = FALSE, wS = 0.5,
eqcoef="mean-mean", robust=FALSE, distribution =
list("normal", par= data.frame(mu =0,sigma=1)))
```

where many of the arguments are defined similarly as those used for `kequate()` (see Sect. 4.5.2 and Table 4.2) with the exception of `P` and `Q`, which in this case are objects of class "matrix" or are objects created by the **R** package **ltm** (Rizopoulos 2006) or **mirt** (Chalmers 2012). These objects contain either the response patterns in population P (or Q) or the estimated IRT model in P (or Q). Quadrature points used for calculating the marginal score distribution are specified using the argument `qpoints`. If no value is provided, then the quadrature points used for the IRT model fitting are used by default. The type of IRT model can be specified using the argument `model`, and the current binary IRT model options are `"2pl"` and `"3pl"` for the 2PL and 3PL models, respectively. When a polytomous IRT model is used, the arguments `catsX`, `catsY`, and `catsA` are used to set the number of response categories for each item in test forms X, Y, and A, respectively.

The `irtose()` function is also capable of giving as output two types of SEE by using the argument `see`. Either analytical or bootstrap standard errors are available by setting `see="bootstrap"` or `see="analytical"`, respectively. In the first case, the argument `replications` is used to set the number of bootstrap replications. When performing equating under the NEAT design, the `eqcoef` argument is used to specify the type of equating coefficients to be used for parameter linking (see Sect. 5.2.1). Current possible options are `eqcoef=c("mean-mean", "mean-gmean", "mean-sigma", "Haebara", "Stocking-Lord")`. Finally, `robust` is a logical indicator of whether a robust covariance matrix (Yuan et al. 2014) should be calculated for the IRT model.

7.3.3 *Examples of IRT Kernel Equating for Binary Scored Items Using kequate*

To exemplify IRT kernel equating for binary-scored data under the NEAT design, we use the admissions data. After the full data sets are loaded for the X and Y tests forms, matrices of binary item responses are stored in the matrix objects `admPneat` and `admQneat`, where the first 80 columns in each matrix correspond to the responses to unique items and the last 40 to common (i.e. anchor) items in both test forms

```
> load("ADMneatX.Rda")
> load("ADMneatY.Rda")
> admPneat0 <- as.matrix(ADMneatX)
> admPneat  <- admPneat0[,c(41:120,1:40)]
> admQneat0 <- as.matrix(ADMneatY)
> admQneat  <- admQneat0[,c(41:120,1:40)]
```

Next, a 2PL IRT model is fitted using the obtained binary matrices as input in the function `ltm()` from the **ltm** package

```
> library(ltm)
> adm2plP <- ltm(admPneat~z1, IRT.param=TRUE)
> adm2plQ <- ltm(admQneat~z1, IRT.param=TRUE)
```

As a final step, the `ltm` objects `adm2plP` and `adm2plQ`, which contain the IRT parameter estimates and their associated covariance matrix, are used as input in `irtose()` to perform the actual IRT kernel equating

```
> library(kequate)
> admNEAT2pl <- irtose("CE", adm2plP, adm2plQ,
+ 0:80, 0:80, 0:40)
```

A summary of the output reads as[2]

```
> summary(admNEAT2pl)
 Design: IRT-OSE CE

 Kernel: gaussian

 Sample Sizes:
   Test X: 2000
   Test Y: 2000

 Score Ranges:
   Test X:
       Min = 0 Max = 80
   Test Y:
       Min = 0 Max = 80
   Test A:
       Min = 0 Max = 40

 Bandwidths Used:
        hxP       hyQ       haP       haQ   hxPlin   hyQlin
        haPlin   haQlin
1 0.6334889 0.6260819 0.6372787 0.6391538 12617.23 13089.33
        6848.809 6868.221

 Equating Function and Standard Errors:
   Score       eqYx     SEEYx
1      0 -0.4087277 0.1338158
2      1  0.4942297 0.1726591
..    .. .......... .........
80    79 79.3247685 0.1953112
81    80 80.2525764 0.1589894

 Comparing the Moments:
          PREAx         PREYa
1    0.0002341643 -0.0005381284
2   -0.0000993506 -0.0027625228
3   -0.0097508659  0.0025430473
4   -0.0207441206  0.0120485135
5   -0.0273741594  0.0222908184
6   -0.0262066529  0.0309160354
7   -0.0151882451  0.0364246762
8    0.0069589351  0.0378498273
9    0.0410743834  0.0345417469
10   0.0877410258  0.0260398591
```

[2]Results for test scores (2, …, 78) have been omitted.

Among other information, the output shows the bandwidth parameters used in the continuization steps, the equated values together with their corresponding SEE, and the value of PRE for the first 10 moments.

We want to emphasize that when performing IRT kernel equating using `irtose()`, other options for the used kernel, such as the logistic kernel and the uniform kernel, as well as the bandwidth selection method (e.g., the rule-based bandwidth selection by adding the argument `altopt=TRUE`) can also be chosen. Further examples showing the use of these alternatives are available on the book's webpage.

7.3.4 Examples of IRT Kernel Equating for Polytomous Scored Items Using kequate

To exemplify IRT kernel equating for polytomous data (Andersson 2016) under the EG design, we use simulated data. Two IRT models for polytomous data are currently available as options of the argument `model` in the `irtose()` function. The graded response model (Samejima 1969) is used by setting `model="grm"`, and the generalized partial credit model (Muraki 1992) is used by setting `model="gpcm"`. In the following example, we simulate data according to a generalized partial credit model.

Assume we have two test forms both containing 10 items that can be scored as 0, 1, or 2. Under this setting, two threshold parameters and one discrimination parameter must be generated for each test item. The following code shows how to obtain randomly generated item parameter data:

```
> Xs <- vector("list", 10)
> Ys <- vector("list", 10)
> set.seed(5)
> for(i in 1:10) Xs[[i]] <- c(rnorm(1, -0.2, 1),
+ rnorm(1, 0.4, 1), runif(1 + 0.2))
> set.seed(7)
> for(i in 1:10) Ys[[i]] <- c(rnorm(1, -0.2, 1),
+ rnorm(1, 0.4, 1), runif(1 + 0.2))
```

The first two entries in each list stored in the objects `Xs` and `Ys` correspond to the threshold parameters, followed by the item discrimination parameter in the third entry. Using these values of item parameters, the `rmvordlogis()` function from the **ltm** package can be used to generate 500 response patterns for both of the test forms as follows:

```
> Xscore <- rmvordlogis(500, Xs, model ="gpcm",
+ z.vals = rnorm(500))
> library(ltm)
> Yscore <- rmvordlogis(500, Ys, model ="gpcm",
+ z.vals = rnorm(500,1,1))
```

Objects `Xscore` and `Yscore` are both 500×10 matrices of item responses that can be modeled using the `mirt()` function from the **mirt** package. The **mirt** package allows the user to model many of the most common IRT models, including the graded response model and the generalized partial credit model. The obtained objects are then read into the `irtose()` function from **kequate** to perform the actual equating. Note that when polytomous data are modeled, it is necessary to specify the number of response categories used for each of the items in the two tests that are equated. This can be done using the arguments `catsX` and `catsY` as shown below

```
> library(mirt)
> X500 <- mirt(Xscore-1, 1, itemtype = rep('gpcm',10),
+ quadpts=61, SE = T)
> Y500 <- mirt(Yscore-1, 1, itemtype = rep('gpcm',10),
+ quadpts=61, SE = T)
> library(kequate)
> EGgpcm <- irtose("EG", X500, Y500, 0:20, 0:20,
+ catsX = rep(3,10), catsY = rep(3,10), model = "GPCM")
```

We can use the regular `get` commands available in **kequate** as described in Sect. 4.6.5 on the obtained object `EGgpcm`. For instance, to retrieve the equated values we write

```
> getEq(EGgpcm)
> getEq(EGgpcm)
 [1]  0.4110581  1.6054127  2.8263102  4.0636845
 [5]  5.3077019  6.5491877  7.7797387  8.9925383
 [9] 10.1819455 11.3426036 12.4692879 13.5573869
[13] 14.6033025 15.6045235 16.5593237 17.4662088
[17] 18.3233796 19.1275365 19.8716110 20.5494851
[21] 21.1591861
```

We point out that it is also possible to perform polytomous IRT kernel equating under a NEAT design using either chained equating or poststratification equating. Further examples of IRT kernel equating under the NEAT design as well as examples assuming the graded response model for the IRT fit can be found on the book's webpage.

7.4 Bayesian Nonparametric Approach to Equating

7.4.1 Bayesian Nonparametric Modeling

Consider the general definition of a parametric statistical model given in Sect. 1.2.2. In the parametric Bayesian framework (e.g., Gelman et al. 2003), a prior $p(\theta)$ is defined on Θ. Parametric Bayesian inference is then based on the posterior distribution $p(\theta \mid x)$, which is proportional to the product of the prior $p(\theta)$ and the likelihood $p(x \mid \theta)$.

The Bayesian nonparametric (BNP) approach (Ghosh and Ramamoorthi 2003; Hjort et al. 2010) starts by focusing on spaces of distribution functions so that uncertainty is expressed on F itself (see Sect. 1.2.2). The prior distribution $p(F)$ is defined on the space $\mathfrak{F}$ of all distribution functions defined on $\mathscr{X}$. If $\mathscr{X}$ is an infinite set, then $\mathfrak{F}$ is infinite-dimensional, and the corresponding prior model $p(F)$ on $\mathfrak{F}$ is termed *nonparametric*. The prior probability model is also referred to as a *random probability measure*, and it essentially corresponds to a distribution on the space of all distributions on the set $\mathscr{X}$. Thus Bayesian nonparametric models are probability models defined on a function space (Müller and Quintana 2004).

7.4.2 BNP Model for Equating

González et al. (2015b) (see also, González et al. 2015a) proposed a Bayesian nonparametric approach for equating that allows the use of covariates in the estimation of the score distribution functions that lead to the equating transformation. The main idea consists of introducing BNP models for a collection of covariate-dependent equating transformations

$$\left\{\varphi_{\mathbf{z}_f,\mathbf{z}_t}(\cdot) : \mathbf{z}_f, \mathbf{z}_t \in \mathscr{Z}\right\}, \tag{7.9}$$

where $\varphi_{\mathbf{z}_f,\mathbf{z}_t}(\cdot) = F^{-1}_{\mathbf{z}_t}(F_{\mathbf{z}_f}(\cdot))$ is the equating function that maps the scores associated with the *from* scale defined by $\mathbf{z}_f$, to the *to* scale related to $\mathbf{z}_t$. Here, $\mathbf{z}_f$ and $\mathbf{z}_t$ are covariate vectors and $\mathscr{Z}$ denotes the covariate space. Assuming data of the form $(t_i, \mathbf{z}_i)$, $i = 1, \ldots, n$, where t_i denotes the test scores and $\mathbf{z}_i$ denotes the covariates, these data are modeled as

$$\begin{aligned} t_i \mid \mathbf{z}_i, F_{\mathbf{z}_i} &\overset{\text{ind}}{\sim} F_{\mathbf{z}_i}, \; i = 1, \ldots, n, \\ \{F_{\mathbf{z}} : \mathbf{z} \in \mathscr{Z}\} &\sim \pi, \end{aligned} \tag{7.10}$$

where π denotes a prior probability measure for $\{F_{\mathbf{z}} : \mathbf{z} \in \mathscr{Z}\}$. Based on the modeling scheme in Eq. (7.10), we are interested in making inferences about the collection of covariate-dependent[3] score distributions $\{F_{\mathbf{z}} : \mathbf{z} \in \mathscr{Z}\}$ that, in turn, will allow us to infer about the family of equating transformations in Eq. (7.9). The computational implementation of the model is based on Markov Chain Monte Carlo (MCMC) methods. In what follows, the presented approach is illustrated using the *single weights* version of the dependent Bernstein polynomial process (DBPP) as the prior probability model so that $\{F_{\mathbf{z}} : \mathbf{z} \in \mathscr{Z}\} \sim w\text{DBPP}(\alpha, \lambda, \Psi, \mathscr{H})$ (Barrientos et al. 2012). More details on the DBPP can be found in Sect. B.9. A BNP model for

[3]In the examples that follow, the dependence is accounted in the form of a regression on the $\mathbf{z}$ covariates.

equating that does not consider the use of covariates in the estimation of the score probability distributions is described in Karabatsos and Walker (2009).

7.4.3 *An Illustration of the BNP Model for Equating in SNSequate*

The Bayesian approach for equating with covariates is implemented in **SNSequate** using the functions: `BNP.eq()` and `BNP.eq.predict()`. The `BNP.eq()` function[4] is used to implement the modeling scheme in Eq. (7.10). A general function call for `BNP.eq()` reads as follows:

```
BNP.eq(scores_x, scores_y, range_scores=NULL,
design="EG", covariates=NULL,...)
```

The arguments `scores_x` and `scores_y` are used to read the score vectors from test forms X and Y, respectively, and the range of values in the score scale is specified using the argument `range_scores`. Currently, the only equating design supported is the EG design. The covariates used for equating are specified in the argument `covariates`. The function gives as output a list of objects that include the samples from the posterior distribution of all the parameters involved in the analysis, and a copy of both the vector of scores and the matrix of covariates used. Also, the object `patterns`, which is described in more detail later, is stored in the list.

The actual prediction of the equating transformation is implemented in the function `BNP.eq.predict()` which receives as input in the argument `model` an object created with `BNP.eq()` and gives as output the equated score values. A general function call reads as follows:

```
BNP.eq.predict(model, from=NULL, into=NULL,...)
```

The arguments `from` and `to` are used to specify the *from* and the *to* scale, defined by the covariates, where the equating is to be performed (see Eq. (7.9)). This arguments can be set using the information provided by the object `patterns` as exemplified below. Also, plot methods have been implemented for objects created using the `BNP.eq.predict()` function that produce graphical displays of the estimated score distributions and the equating transformation.

To illustrate how these functions are used to perform a Bayesian equating with covariates we use the `SEPA` data. In the original data, test forms X and Y are administered to 1,458 and 2,619 test takers, respectively. Here, we use a random sample of 50 test takers for each form. The following code is used to read in the data and to obtain the random samples

---

[4]The `BNP.eq()` function makes use of other functions from the **R** package **DPpackage** (Jara et al. 2011), which is also downloaded and installed when installing **SNSequate**.

```
> load("SEPA")
> set.seed(6)
> sample_size <- 50
> idx_x <- sample(length(scores_x), sample_size)
> idx_y <- sample(length(scores_y), sample_size)
> scores_x <- scores_x[idx_x]
> scores_y <- scores_y[idx_y]
> Z <- Z[c(idx_x, idx_y+length(scores_x)), ]
```

The objects `scores_x` and `scores_y` are vectors of length 50 that contain the scores in test forms X and Y, respectively. The covariates values are stored in the object `Z` and they are coded as male (`M`) or female (`F`) for gender, and `MUN`, `PS`, `PP` for school type (e.g., municipal, subsidized, private). The MCMC scheme is run using the following code[5]

```
> mod.bnp <- BNP.eq(scores_x, scores_y,
+ range_scores = c(0, 50), covariates=Z)
```

As mentioned before, the information provided by the object `patterns` is used to specify the *from* and the *to* scale where the equating is to be performed. The object appearance is as follows

```
> mod.bnp$patterns
   x..Intercept. x.FormY x.GenderM x.SchoolPP x.SchoolPS
1              1       0         0          0          0
2              1       0         0          0          1
3              1       0         0          1          0
4              1       0         1          0          0
5              1       0         1          0          1
6              1       0         1          1          0
7              1       1         0          0          0
8              1       1         0          0          1
9              1       1         0          1          0
10             1       1         1          0          0
11             1       1         1          0          1
12             1       1         1          1          0
```

The object is a design matrix where each row pattern represent a combination of covariates defining a particular score distribution. For instance, the first pattern defines the distribution of X scores for females attending municipal schools, whereas the eleventh defines the distribution of Y scores for males attending subsidized schools, and so on. Because in this case gender and test form both have two levels and school type has three levels, a total of $2 \times 2 \times 3 = 12$ different combinations are possible. The BNP equating model can thus be used to equate forms administered on different groups implied by covariates values (customized equating).

Suppose we want to equate scores on test form X to scores on test form Y. Then we need to write[6]

[5]Running the `BNP.eq()` function can take a considerable amount of computation time.

[6]Note that the covariates that are not directly involved in the equating of interest are automatically integrated out by the function.

```
> equ <- BNP.eq.predict(mod.bnp, from=c(1,2,3,4,5,6),
+ into=c(7,8,9,10,11,12))
```

The output shows the score scale and the equated values and can be obtained as (score values between 4 and 47 are omitted)

```
> equ
Equated values:

 Score   eYx
     0  0.00
     1  1.89
     2  2.95
     3  3.98
    .......
    48 49.37
    49 49.66
    50 50.00
```

A graphical display of the estimated equating transformation is obtained by writing `plot(equ)`. Plots of the estimated score distributions and the corresponding CDFs of scores X and Y can be obtained adding the arguments `which` and `what`. For instance, the probability density function (PDF) of X scores is obtained by writing

```
> plot(equ, which="X", what="PDF")
```

As a final example, we reproduce some of the results shown in González et al. (2015b). Figure 7.2 shows a graphical display of the estimated score distributions as well as equating transformations for the case when only the type of school is considered as a covariate in the model. In this figure panels (c), (f) and (i) show also the equating function estimated when no covariates are considered in the analysis (dashed line). It can be seen that the equating function appears to depend on the school type. When school type is included in the estimation, the equating functions show departure from the reference. The reader can verify that almost no departures from the reference equating function (i.e., the one which does not include covariates) occur when considering only the gender covariate. Thus, we can conclude that the equating function appears to depend only on the school type but not on the gender. The **R** code necessary to produce Fig. 7.2 and verify these findings can be obtained from the book's webpage.

7.5 Assessing the Equating Transformation

This section is based on the ideas in Wiberg and González (2016). Instead of the usual practice of targeting different parts of the equating process and aiming to evaluate the process from different aspects, we propose a new perspective in which the equating transformation should be seen as a standard estimator that

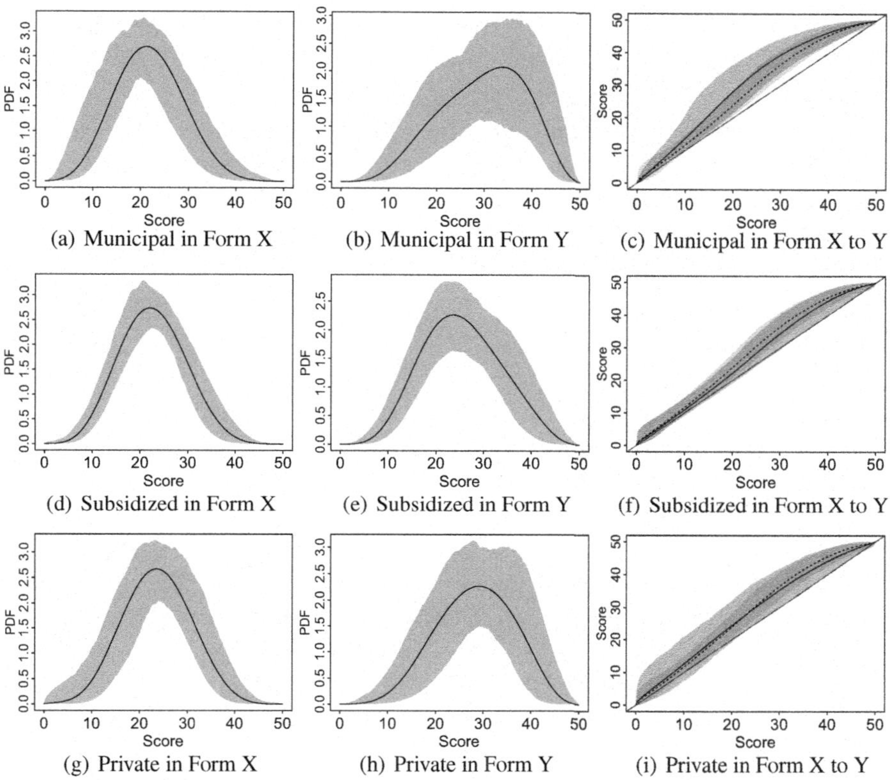

Fig. 7.2 SEPA data – Posterior mean (*continuous line*), and 95% point-wise HPD intervals (*in gray*) for: conditional (school type and test form) densities (panels (**a**), (**b**), (**d**), (**e**), (**g**), (**h**)), and equating functions (panels (**c**), (**f**), and (**i**)). The *dotted* and *dashed line* displayed in (panel (**c**), (**f**), and (**i**)) corresponds to the identity function and the posterior mean of the equating function from test form X to Y (i.e. ignoring covariates), respectively

needs to be statistically assessed. The usual practice makes use of *equating-specific evaluation measures*, e.g., PRE in kernel equating (described in Sect. 4.6.3), the SEE (described in Sect. 4.6.1), and the difference that matters (DTM, Dorans and Feigenbaum 1994). Among these measures, there are also summary indices (Harris and Crouse 1993) that have been used for evaluations in equating, although not in the way we assess equating transformations here. In contrast, the new perspective makes use of general statistical evaluation measures such as bias, standard errors (SE), mean squared error (MSE), and root mean squared error (RMSE), all of which are built on the general statistical perspective described in Chap. 1. Note, however, that in order to evaluate these statistical measures in practice, we need to use the Monte Carlo method with replicated data generated from a known probability model. The formal definitions and details of the way these measures are calculated in practice when equating test scores are given in Sect. B.10. In what follows, we use the kernel equating framework to illustrate the statistical assessment of an equating transformation.

7.5.1 An Illustration of Assessing $\hat{\varphi}(x)$ in Kernel Equating Using SNSequate

Wiberg and González (2016) used the `Math20EG` data to compare a logistic and a Gaussian kernel using all of the described measures of assessment. In this section, we will reproduce some of the results of that study.

In order to make a fair comparison, *two* sets of true equated values – one obtained using a logistic kernel and the other one using a Gaussian kernel- are generated as follows:

```
> data(Math20EG)
> mod.gauss.true <- ker.eq(scores=Math20EG,
+ kert="gauss", hx=NULL,hy=NULL,degree=c(2,3),
+ design="EG")
> mod.logis.true <- ker.eq(scores=Math20EG,
+ kert="logis", hx=NULL,hy=NULL,degree=c(2,3),
+ design="EG")
```

Both objects `mod.gauss.true` and `mod.logis.true` actually contain not only the corresponding equated values but also the associated SEE, the estimated score probabilities, and the PRE values, which can be retrieved as follows:

```
> true.r <- mod.gauss.true$rj
> true.s <- mod.gauss.true$sk
> true.eq.gauss <- mod.gauss.true$eqYx
> true.eq.logis <- mod.logis.true$eqYx
> true.see.gauss <- mod.gauss.true$SEEYx
> true.see.logis <- mod.logis.true$SEEYx
> true.pre.gauss <- PREp(mod.gauss.true,10)$preYx
> true.pre.logis <- PREp(mod.logis.true,10)$preYx
```

To make a fair assessment of the equating transformation, we need to simulate the following four possible scenarios:

1. True equated values are from a Gaussian kernel model and equated scores are estimated using a Gaussian kernel.
2. True equated values are from a Gaussian kernel model and equated scores are estimated using a logistic kernel.
3. True equated values are from a logistic kernel model and equated scores are estimated using a Gaussian kernel.
4. True equated values are from a logistic kernel model and equated scores are estimated using a logistic kernel.

In order to use the statistical evaluation methods in practice, a large set of replications is performed as described in Sect. B.10. Score frequencies are generated using a multinomial probability model with parameters stored in the objects `true.r` and `true.s`. Then, for each replication, values of equated scores, SEE, and PRE are stored in the objects `res.gauss`, `res.logis`, `resSEE.gauss`, `resSEE.logis`, `resPRE.gauss`, `resPRE.logis`, respectively. The following code shows how these tasks are carried out, starting by setting the number of replications and the storage space.

```
> sims = 1000
> res.X <-matrix(0,nrow=sims,ncol=21)
> res.Y <-matrix(0,nrow=sims,ncol=21)
> res.gauss <-matrix(0,nrow=sims,ncol=21)
> res.logis <-matrix(0,nrow=sims,ncol=21)
> resSEE.gauss <-matrix(0,nrow=sims,ncol=21)
> resSEE.logis <-matrix(0,nrow=sims,ncol=21)
> resPRE.gauss <-matrix(0,nrow=sims,ncol=10)
> resPRE.logis <-matrix(0,nrow=sims,ncol=10)
```

```
> i=0
> it=0
> while(it<sims){
+ i=i+1
+ X<-rmultinom(1, 1453, true.r)
+ Y<-rmultinom(1, 1455, true.s)
+ mod.gauss <-tryCatch(ker.eq(scores=cbind(X,Y),
+ kert="gauss",hx=NULL,hy=NULL,degree=c(2,3),
+ design="EG"),error=function(e)NULL)
+ mod.logis <-tryCatch(ker.eq(scores=cbind(X,Y),
+ kert="logis",hx=NULL,hy=NULL,degree=c(2,3),
+ design="EG"),error=function(e)NULL)
+
+ if(!any(lapply(list(mod.gauss,mod.logis),is.null))){
+ it=it+1
+ res.X[it,]<-X
+ res.Y[it,]<-Y
+ res.gauss[it,]<-mod.gauss$eqYx
+ res.logis[it,]<-mod.logis$eqYx
+ resSEE.gauss[it,]<-mod.gauss$SEEYx
+ resSEE.logis[it,]<-mod.logis$SEEYx
+ resPRE.gauss[it,]<-PREp(mod.gauss,10)$preYx
+ resPRE.logis[it,]<-PREp(mod.logis,10)$preYx }
+ print(i)
+ print(it)
+}
```

Using the replications, we can calculate the different statistical measures defined in Eqs. (B.33), (B.34), (B.35), and (B.36). For instance, we can create a bias function as

```
> Bias <- function(results,parameter){
+ rel.bias<-matrix(0,nrow=nrow(results),ncol=ncol(results))
+   for(i in 1:nrow(results)){
+          rel.bias[i,]<-(results[i,]-parameter)}
+   res <- apply(rel.bias,2,mean)
+   return(res)}
```

and we can visualize the results for the four different scenarios as shown in Fig. 7.3. Note that because we are using simulations the appearance of the bias might slightly change if the analysis is redone. The following code was used to produce the figure:

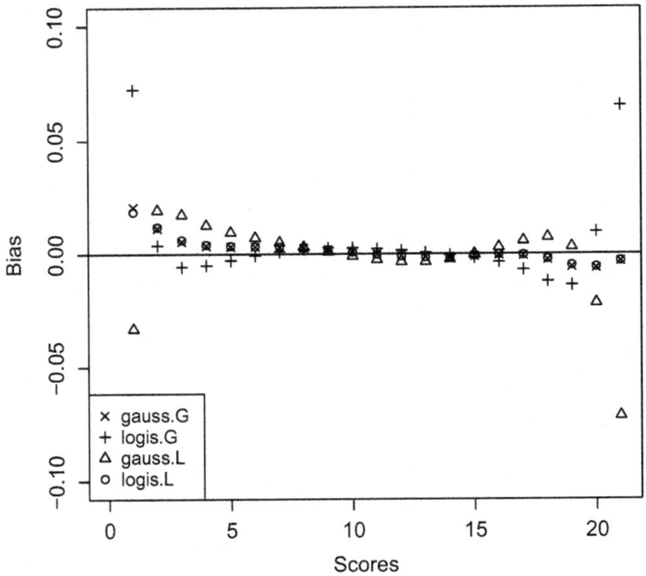

Fig. 7.3 Bias for the four examined scenarios

```
> b.gauss.G <- Bias(res.gauss,true.eq.gauss)
> b.logis.G <- Bias(res.logis,true.eq.gauss)
> b.gauss.L <- Bias(res.gauss,true.eq.logis)
> b.logis.L <- Bias(res.logis,true.eq.logis)
> plot(b.gauss.G,ylim=c(-0.1,0.1),pch=4,
+ ylab="bias",xlab="Scores")
> abline(h=0)
> points(b.logis.G,pch=3)
> points(b.gauss.L,pch=2)
> points(b.logis.L,pch=1)
> legend("bottomleft",pch=c(4,3,2,1)
+ c("gauss.G","logis.G","gauss.L","logis.L"))
```

It can be seen that the bias differs between the examined scenarios, especially at the end points. To further examine the equating transformation, it is possible to write functions and create plots for the MSE, RMSE and SE in a similar way as shown here. Note that besides using these measures it is also possible to use summary indices of equating specific measures, although they need to be adjusted to be used with replication as described in Wiberg and González (2016).

7.6 Summary

In this chapter, some of the recent developments in equating research have been briefly summarized and exemplified using the **R** packages **kequate** and **SNSequate**. Research about equating has exploded in recent years, and thus there are several other topics that could have been incorporated here. We did, however, restrict the contents of this chapter to some of the methods that have been implemented in **R** packages.

With many new equating methods available, it is more and more important to have trustworthy and efficient methods to evaluate the equating transformations. In this chapter we have emphasized the importance of viewing the equating transformation as a statistical estimator and have proposed a general strategy for the statistical assessment of equating transformations.

The rapid changes in the test equating world suggest that it is of great importance to facilitate building on the most recent research. With this book, we want to strongly encourage people to use **R** to implement new methods and functions that are compiled into **R** packages that are available for everyone to use. In this way, the next generation of researchers will be able to reach further and will be able to do so much more quickly than their predecessors.

References

Andersson, B. (2016). Asymptotic standard errors of observed-score equating with polytomous IRT models. *Journal of Educational Measurement, 53*(4), 459–477.

Andersson, B., Bränberg, K., & Wiberg, M. (2013). Performing the kernel method of test equating with the package kequate. *Journal of Statistical Software, 55*(6), 1–25.

Andersson, B., & von Davier, A. A. (2014). Improving the bandwidth selection in kernel equating. *Journal of Educational Measurement, 51*(3), 223–238.

Andersson, B., & Wiberg, M. (2014). *IRT observed-score kernel equating with the R package kequate*. R: Vignette. Retrieved from https://cran.r-project.org/web/packages/kequate/vignettes/irtguide.pdf

Andersson, B., & Wiberg, M. (2016). Item response theory observed-score kernel equating. *Psychometrika*. doi: 10.1007/s11336–016–9528–7.

Barrientos, A. F., Jara, A., & Quintana, F. A. (2012). On the support of MacEachern's dependent Dirichlet processes and extensions. *Bayesian Analysis, 7*, 277–310.

Chalmers, R. P. (2012). Mirt: A multidimensional item response theory package for the R environment. *Journal of Statistical Software, 48*(6), 1–29.

Cid, J. A., & von Davier, A. A. (2015). Examining potential boundary bias effects in kernel smoothing on equating: An introduction for the adaptive and Epanechnikov kernels. *Applied Psychological Measurement, 39*(3), 208–222.

Dorans, N., & Feigenbaum, M. (1994). Equating issues engendered by changes to the SAT and PSAT/NMSQT. *Technical issues related to the introduction of the new SAT and PSAT/NMSQT* (pp. 91–122).

Epanechnikov, V. (1969). Non-parametric estimation of a multivariate probability density. *Theory of Probability and Its Applications, 14*, 153–158.

Gelman, A., Carlin, J., Stern, H., & Rubin, D. (2003). *Bayesian data analysis* (2nd ed.). Boca Raton: Chapman and Hall.

Ghosh, J., & Ramamoorthi, R. (2003). *Bayesian nonparametrics*. New York: Springer.

González, J. (2014). SNSequate: Standard and nonstandard statistical models and methods for test equating. *Journal of Statistical Software, 59*(7), 1–30.

González, J., Barrientos, A. F., & Quintana, F. A. (2015a). A dependent Bayesian nonparametric model for test equating. In R. Millsap, D. Bolt, L. van der Ark, & W.-C. Wang (Eds.), *Quantitative psychology research* (pp. 213–226). Cham: Springer International Publishing.

González, J., Barrientos, A. F., & Quintana, F. A. (2015b). Bayesian nonparametric estimation of test equating functions with covariates. *Computational Statistics & Data Analysis, 89*, 222–244.

González, J., & von Davier, A. A. (2017). An illustration of the Epanechnikov and adaptive continuization methods in kernel equating. In L. A. van der Ark, M. Wiberg, S. A. Culpeppe, J. A. Douglas, & W.-C. Wang (Eds.), *Quantitative psychology – 81st annual meeting of the psychometric society, Asheville, North Carolina, 2016*. New York: Springer.

Häggström, J., & Wiberg, M. (2014). Optimal bandwidth selection in observed-score kernel equating. *Journal of Educational Measurement, 51*(2), 201–211.

Hall, P., Marron, J., & Park, B. U. (1992). Smoothed cross-validation. *Probability Theory and Related Fields, 92*(1), 1–20.

Harris, D. J., & Crouse, J. D. (1993). A study of criteria used in equating. *Applied Measurement in Education, 6*(3), 195–240.

Hjort, N. L., Holmes, C., Müller, P., & Walker, S. (2010). *Bayesian nonparametrics*. Cambridge: Cambridge University Press.

Jara, A., Hanson, T., Quintana, F., Müller, P., & Rosner, G. (2011). Dppackage: Bayesian non-and semi-parametric modelling in R. *Journal of Statistical Software, 40*, 1–30.

Karabatsos, G., & Walker, S. (2009). A Bayesian nonparametric approach to test equating. *Psychometrika, 74*(2), 211–232.

Liang, T., & von Davier, A. A. (2014). Cross-validation: An alternative bandwidth-selection method in kernel equating. *Applied Psychological Measurement, 38*(4), 281–295.

Lord, F., & Wingersky, M. (1984). Comparison of IRT true-score and equipercentile observed-score "equatings". *Applied Psychological Measurement, 8*(4), 453–461.

Müller, P., & Quintana, F. (2004). Nonparametric Bayesian data analysis. *Statistical Science, 19*, 95–110.

Muraki, E. (1992). A generalized partial credit model: Application of an EM algorithm. *Applied Psychological Measurement, 16*, 159–176.

Rizopoulos, D. (2006). ltm: An R package for latent variable modeling and item response theory analyses. *Journal of Statistical Software, 17*(5), 1–25.

Samejima, F. (1969). *Estimation of latent ability using a response pattern of graded scores. (Psychometrika Monograph No. 17)*. Richmond, VA: Psychometric Society.

Silverman, B. (1986). *Density estimation for statistics and data analysis* (Vol. 3). London: Chapman and Hall.

Thissen, D., Pommerich, M., Billeaud, K., & Williams, V. S. L. (1995). Item response theory for scores on tests including polytomous items with ordered responses. *Applied Psychological Measurement, 19*(1), 39–49.

Wiberg, M., & González, J. (2016). Statistical assessment of estimated transformations in observed-score equating. *Journal of Educational Measurement, 53*(1), 106–125.

Yuan, K.-H., Cheng, Y., & Patton, J. (2014). Information matrices and standard errors for MLEs of item parameters in IRT. *Psychometrika, 79*(2), 232–254.

Appendix A
Installing and Reading Data in R

A.1 Installing R

R (R Core Team 2016) is a free software environment that runs on Windows, MacOS and Unix platforms. **R** can be downloaded from the Comprehensive R Archive Network (CRAN) webpage[1]: http://cran.r-project.org/.

There are precompiled binary distributions of the base system and contributed packages for different platforms. The following list shows the links where these distributions can be downloaded for different platforms when installing **R** for the first time:

1. **R** for Linux (http://cran.r-project.org/bin/linux/)
2. **R** for MacOS (http://cran.r-project.org/bin/macosx/)
3. **R** for Windows (http://cran.r-project.org/bin/windows/)

All of the examples given in this book have been created and tested on Windows using **R** version 3.3.2. To install **R**, run the executable file (R-3.3.2-win.exe) by double clicking on it, which will start the **R** for Windows setup wizard.

A.1.1 R Studio

After installing the **R** software it is possible to install **R studio** which is a graphical user interface. **R studio** (http://www.rstudio.org/) provides the usual **R** console and has the advantage of providing a history/workspace window with common data set features such as load/save/import. There are also a window with easy access to files

[1]It is recommended to use a mirror site near the place where the program is being downloaded. A full list of mirrors can be found at https://cran.r-project.org/mirrors.html.

J. González, M. Wiberg, *Applying Test Equating Methods*,
Methodology of Educational Measurement and Assessment,
DOI 10.1007/978-3-319-51824-4

and list of available packages. The Plots tab include easy export to a PDF file, a GIF image and copy features to facilitate to insert plots into Word documents.

A.2 Installing and Loading R Packages

Once **R** has been installed, it is time to open the program and start working. To install contributed **R** packages that do not come with the initial distribution, one can go to the tab "Packages/Install package(s)". Choose the package of interest (e.g. **equate**) and press ok to install the package. To use the package, one can either go to the menu and select "Packages/load package" or simply type in the **R** command prompt

```
> library(equate)
```

Using the library command gives access to all functions within that package. When help is needed with an **R** function or an **R** package, the user can simply write `help()` in the **R** command prompt. For instance, if help is needed on the use of the `equate()` function, one can write

```
> help(equate)
```

It should be noted that there are a number of books, online help resources, and online books (simply search for "R Statistics book") where information about **R** can be obtained.

The **R** packages used for the examples shown in this book have been executed under the following versions:

- **equate** version 2.0.6
- **SNSequate** version 1.2.2
- **kequate** version 1.6.0
- **equateIRT** version 2.0.1
- **ltm** version 1.0.0
- **mirt** version 1.20.1

A.3 Working Directory and Accessing Data

It is convenient to choose a working directory and to store everything connected to a program in such a folder. This can be done in two ways, either by right clicking on the **R** icon on the desktop and setting which directory **R** always should start in, or from inside **R**. In the **R** command prompt, the current working directory can be identified by writing

```
> getwd()
```

and to set a working directory we can write

```
> setwd("C:/Users/documents/MyRProject")
```

A.4 Loading Data of Different File Formats

In Chap. 2 we described how the data sets used in this book can be loaded. If the user has data with different file formats, it is possible to use other functions such as `read.table()`. Assume that the `ADM` data described in Chap. 2 are in another file format than .Rda. The following examples show how data with different formats can be read into **R**. The data sets used in the examples can be download from the book's webpage.

First, if we want to read in a text file (.txt), that is separated with commas and has headers, we can write the following

```
> ADMt<-read.table(file="ADMt.txt",sep="\t",header=TRUE)
```

If the data are in an Excel file, an easy way to read the data into **R** is to save the file as either a .dat file or a .csv file. The following commands can be used after saving the files as either of those file types

```
> ADMc1 <-read.csv("ADMc.csv",sep=";",header=TRUE)
> ADMc2 <-read.table("ADMc.csv",sep=";",header=TRUE)
> ADMd  <-read.table("ADMd.dat",header=T)
```

The same kind of data can be read directly from the Excel file (.xls) if the **R** package **XLConnect** is installed. In this example "Sheet1" in the Excel file is named "ADMsheet‘". Reading the data into **R** can be done by writing

```
> library(XLConnect)
> ADMe   <- loadWorkbook("ADMe.xls")
> ADMee <- readWorksheet(ADMe, sheet="ADMsheet")
```

Finally, we can read in an SPSS file (.sav) using the **R** package **foreign**

```
> library(foreign)
> ADMs <- read.spss("ADMs.sav",to.data.frame=TRUE)
```

We can use any of these files with the **R** package **equate** to obtain the sum scores for the verbal section of the admissions test as described in Chap. 2. Each line of the following code yields the same result as the object `"verb.y"` in Sect. 2.2.4

```
> ADMtSum   <- apply(ADMt[,41:120],1,sum)
> ADMc1Sum <- apply(ADMc1[,41:120],1,sum)
> ADMc2Sum <- apply(ADMc2[,41:120],1,sum)
> ADMdSum   <- apply(ADMd[,41:120],1,sum)
> ADMeSum   <- apply(ADMee[,41:120],1,sum)
> ADMsSum   <- apply(ADMs[,41:120],1,sum)
```

One way to assure that data are read correctly into **R** is to visualize the data set. To perform a visual inspection, we can, for example, write

```
> View(ADMt)
```

Reference

R Core Team (2016). *R: A language and environment for statistical computing*. Vienna, Austria: R Foundation for Statistical Computing.

Appendix B
Additional Material

In this appendix we give more details on definitions, derivations, and other relevant material covered throughout the chapters of the book.

B.1 Design Functions

As seen in Sect. 1.3.1, score data can be modeled separately (EG design) or jointly (SG, CB, NEAT, or NEC designs). The design function **DF** is used to map the score probabilities (either univariate or bivariate) into **r** and **s**. Thus, different design functions (DF) are used for different equating designs (von Davier et al. 2004). For instance, under the EG design, because two independent vectors of scores are obtained, the DF is just the identity function, i.e.,

$$\begin{pmatrix} \mathbf{r} \\ \mathbf{s} \end{pmatrix} = \mathbf{DF}(\mathbf{r}, \mathbf{s}) = \begin{pmatrix} \mathbf{I}_J & 0 \\ 0 & \mathbf{I}_K \end{pmatrix} \begin{pmatrix} \mathbf{r} \\ \mathbf{s} \end{pmatrix} \tag{B.1}$$

The situation is different, however, when bivariate distributions are considered (e.g., under the SG, CB, NEAT, and NEC designs). Define **P** as the $J \times K$ matrix with entries $p_{jk} = \Pr(X = x_j, Y = y_k)$. Further, define the vectorized version of **P** as

$$v(\mathbf{P}) = \begin{pmatrix} \mathbf{p}_1 \\ \vdots \\ \mathbf{p}_K \end{pmatrix}, \tag{B.2}$$

J. González, M. Wiberg, *Applying Test Equating Methods*,
Methodology of Educational Measurement and Assessment,
DOI 10.1007/978-3-319-51824-4

where $\mathbf{p}_k$ is the k-th column of $\mathbf{P}$. In the SG design, the DF is defined as

$$\begin{pmatrix} \mathbf{r} \\ \mathbf{s} \end{pmatrix} = \mathbf{DF}(\mathbf{P}) = \begin{pmatrix} \mathbf{M} \\ \mathbf{N} \end{pmatrix} v(\mathbf{P}), \tag{B.3}$$

where $\mathbf{M} = (\mathbf{I}_J \cdots \mathbf{I}_J)$ is a $J \times KJ$ matrix and

$$\mathbf{N} = \begin{pmatrix} \mathbf{1}_J^t & \mathbf{0}_J^t & \cdots & \mathbf{0}_J^t \\ \vdots & \vdots & \vdots & \vdots \\ \mathbf{0}_J^t & \cdots\cdots & \mathbf{0}_J^t & \mathbf{1}_J^t \end{pmatrix} \tag{B.4}$$

is a $K \times KJ$ matrix.

The CB design can be obtained by using two independent SG designs. In this case, the DF is defined as

$$\begin{pmatrix} \mathbf{r} \\ \mathbf{s} \end{pmatrix} = \mathbf{DF}(\mathbf{P}_{(12)}, \mathbf{P}_{(21)}) = \begin{pmatrix} w_X\mathbf{M} & (1-w_X)\mathbf{M} \\ (1-w_Y)\mathbf{N} & w_Y\mathbf{N} \end{pmatrix} \begin{pmatrix} v(\mathbf{P}_{(12)}) \\ v(\mathbf{P}_{(21)}) \end{pmatrix} \tag{B.5}$$

where the weights w_X and w_Y satisfies $0 \leq w_X, w_Y \leq 1$.

DFs are different under the NEAT design depending on whether CE or PSE is used. For CE, the DF is defined as

$$\begin{pmatrix} \mathbf{r}_P \\ \mathbf{t}_P \\ \mathbf{t}_Q \\ \mathbf{s}_Q \end{pmatrix} = \mathbf{DF}(\mathbf{P}, \mathbf{Q}) = \begin{pmatrix} \mathbf{DF}_P(\mathbf{P}) \\ \mathbf{DF}_Q(\mathbf{Q}) \end{pmatrix} = \begin{pmatrix} \begin{pmatrix} \mathbf{M}_P \\ \mathbf{N}_P \end{pmatrix} & \mathbf{0} \\ \mathbf{0} & \begin{pmatrix} \mathbf{N}_Q \\ \mathbf{M}_Q \end{pmatrix} \end{pmatrix} \begin{pmatrix} v(\mathbf{P}) \\ v(\mathbf{Q}) \end{pmatrix} \tag{B.6}$$

The DF for NEAT PSE is given by

$$\begin{pmatrix} \mathbf{r} \\ \mathbf{s} \end{pmatrix} = \mathbf{DF}(\mathbf{P}, \mathbf{Q}) = \begin{pmatrix} \sum_l \left(w + \frac{(1-w)\sum_k q_{kl}}{\sum_j p_{jl}} \right) \mathbf{p}_l \\ \sum_l \left((1-w) + \frac{w \sum_j p_{jl}}{\sum_k q_{kl}} \right) \mathbf{q}_l \end{pmatrix} \tag{B.7}$$

where $\mathbf{p}_l = (p_{1l}, p_{2l}, \ldots, p_{Jl})^t$ and $\mathbf{q}_l = (q_{1l}, q_{2l}, \ldots, q_{Kl})^t$. Finally, the DF for the NEC design (Wiberg and Bränberg 2015) is given by

$$\begin{pmatrix} \mathbf{r} \\ \mathbf{s} \end{pmatrix} = \mathbf{DF}(\mathbf{P}, \mathbf{Q}) = \begin{pmatrix} \sum_m \left(w + \frac{(1-w)t_{Qm}}{t_{Pm}} \right) \mathbf{p}_m \\ \sum_m \left((1-w) + \frac{w t_{Pm}}{t_{Qm}} \right) \mathbf{q}_m \end{pmatrix}, \tag{B.8}$$

where p_m is the mth column of $\mathbf{P}$ and q_m is the mth column of $\mathbf{Q}$ and the summation is over all the m possible combinations of the values of the covariates, in other words a summation over all columns in the matrices.

B.2 C Matrices

The polynomial log-linear models described in Sect. 2.3.1 can be written in matrix form as follows

$$log(\mathbf{r}) = \alpha + \mathbf{u} + \mathbf{B}^t\boldsymbol{\beta}$$

If $\hat{\boldsymbol{\beta}}$ is the maximum likelihood estimator (MLE) of $\boldsymbol{\beta}$, the estimator of r_j is $\hat{r}_j = \hat{r}_j(\boldsymbol{\beta})$, i.e. the MLE or fitted value of r_j. It is assumed that the estimators, $\hat{\mathbf{r}}$ and $\hat{\mathbf{s}}$, are obtained separately (independently) so that

$$Cov(\hat{\mathbf{r}}, \hat{\mathbf{s}}) = \Sigma_{\hat{r},\hat{s}} = \mathbf{0}.$$

The following theorem establishes a computationally easy way to calculate $\Sigma_{\hat{r},\hat{s}}$.

Theorem *If* $\hat{\mathbf{r}}$ *is the MLE of a log-linear model for* $\mathbf{r}$, *the estimated covariance matrix* $\Sigma_{\hat{r}} = Cov(\mathbf{r})$ *can be obtained as*

$$\Sigma_{\hat{r}} = \mathbf{C}_r\mathbf{C}_r^t\ ,$$

where $\mathbf{C}_r$ *is a* $J \times T_r$ *matrix*

$$\mathbf{C}_r = N^{-1/2}\mathbf{D}_{\sqrt{r}}\mathbf{Q}.$$

The diagonal matrix, $\mathbf{D}_{\sqrt{r}}$ *has elements* $\sqrt{\hat{r}_j}$. *Also,* $\mathbf{Q}$ *is a* $J\times T_r$ *orthogonal matrix that comes from the following* **QR** *factorization*

$$[\mathbf{D}_{\sqrt{r}} - \sqrt{\hat{\mathbf{r}}}\hat{\mathbf{r}}^t]\mathbf{B}^t = \mathbf{QR}\ ,$$

where $\mathbf{Q}$ *is a* $J \times T_r$ *matrix with orthogonal columns and* $\mathbf{R}$ *is an* $T_r \times T_r$ *upper triangular matrix. A proof and more details can be seen in Holland and Thayer (1989) and von Davier et al. (2004).*

B.3 Calculation of the SEE

To calculate the SEE, start by letting $\mathbf{R}$ and $\mathbf{S}$ be the vectors of the pre-smoothed score distributions obtained under any of the data collection designs. If we assume that they are estimated independently, we can write the covariances as (see Sect. B.2)

$$\mathrm{Cov}\begin{pmatrix}\hat{\mathbf{R}}\\ \hat{\mathbf{S}}\end{pmatrix} = \begin{pmatrix}\mathbf{C}_R\mathbf{C}_R^t & 0\\ 0 & \mathbf{C}_S\mathbf{C}_S^t\end{pmatrix} = \mathbf{CC}^t\ , \tag{B.9}$$

where

$$\mathbf{C} = \begin{pmatrix} \mathbf{C}_R & 0 \\ 0 & \mathbf{C}_S \end{pmatrix}. \tag{B.10}$$

The pre-smoothed score distributions, obtained in kernel equating (step 1) can then be transformed into $\mathbf{r}$ and $\mathbf{s}$ through the DF given in Sect. B.1 for the different designs. The Jacobian of the DF is defined as

$$\mathbf{J}_{DF} = \begin{pmatrix} \frac{\partial \mathbf{r}}{\partial \mathbf{R}} & \frac{\partial \mathbf{r}}{\partial \mathbf{S}} \\ \frac{\partial \mathbf{s}}{\partial \mathbf{R}} & \frac{\partial \mathbf{s}}{\partial \mathbf{S}} \end{pmatrix}. \tag{B.11}$$

Further, the Jacobian of the equating transformation in Eq. (4.10) is defined by

$$\mathbf{J}_{\varphi} = \left(\frac{\partial \varphi}{\partial \mathbf{r}}, \frac{\partial \varphi}{\partial \mathbf{s}} \right). \tag{B.12}$$

Note, if $\hat{\mathbf{R}}$ and $\hat{\mathbf{S}}$ are approximately normal distributed with mean $\mathbf{R}$ and $\mathbf{S}$, respectively, and variances given in Eq. (B.9), then

$$\mathrm{SEE}_Y(x) = \left\| \mathbf{J}_{\varphi} \mathbf{J}_{DF} \mathbf{C} \right\| \tag{B.13}$$

where $\|\boldsymbol{v}\|$ is the Euclidian norm of the vector $\boldsymbol{v}$. For more details refer, to von Davier et al. (2004).

B.4 Score Distributions Under the NEAT Design

To formally derive the score distributions of X and Y on T let us use $f_{XP}(x \mid a)$ and $f_{XQ}(x \mid a)$ to denote the conditional distributions of X scores on P and Q, respectively. Similarly, $f_{YP}(y \mid a)$ and $f_{YQ}(y \mid a)$ are the conditional distributions for Y scores on P and Q, respectively. Further, because both samples (and thus both populations) take the anchor test A, we define $f_{AP}(a)$ and $f_{AQ}(a)$ as the (marginal) distributions for anchor scores in P and Q, respectively. Following the definition of a synthetic population, we have that

$$f_{AT}(a) = w_P f_{AP}(a) + w_Q f_{AQ}(a),$$

and thus the score distributions of X and Y on T can be obtained as

$$f_{XT}(x) = \left[\int f_{XP}(x \mid a) f_{AP}(a) da \right] w_P + \left[\int f_{XQ}(x \mid a) f_{AQ}(a) da \right] w_Q \tag{B.14}$$

$$f_{YT}(y) = \left[\int f_{YP}(y \mid a) f_{AP}(a) da \right] w_P + \left[\int f_{YQ}(y \mid a) f_{AQ}(a) da \right] w_Q \tag{B.15}$$

Because under a NEAT design test form X is only administered to population P and test form Y is only administered to population Q, the conditional distributions $f_{XQ}(x \mid a)$ and $f_{YP}(y \mid a)$ cannot be estimated from the collected data.[1] However, note that if we assume that the conditional distributions of $X \mid A$ are the same in both populations P and Q and analogously for that of $Y \mid A$, it can easily be shown that Eqs. (B.14) and (B.15) become

$$f_{XT}(x) = \int f_{XP}(x \mid a) f_{AT}(a) da$$

$$f_{YT}(y) = \int f_{YQ}(y \mid a) f_{AT}(a) da$$

which can be estimated from the observed quantities. The CDFs $F_{XT}(x)$ and $F_{YT}(y)$ can be obtained by cumulating $f_{XT}(x)$ and $f_{YT}(y)$ over values of X and Y, respectively. With this quantities, the equating transformation $\varphi_T(x) = F_{YT}^{-1}(F_{XT}(x))$ is obtained.

A different approach for the NEAT design that does not make use of a synthetic population for equating can be found in San Martín and González (2017).

B.5 The Lord-Wingersky Algorithm

Let $X = \sum_{j=1}^{J} X_{ij}$ be the sum score of a test taker with ability θ. If we assume that test takers with a given ability θ correctly answer each of J items of a test with probability p_j $(i = 1, \ldots, J)$, then the conditional distribution of X for a given θ is called the *compound binomial* (Lord and Novick 1968) and is defined as

$$\Pr(X = x_j \mid \theta) = f(x \mid \theta) = \sum_{\sum x_j = x} \left(\prod_{j=1}^{J} p_j^{x_j} q_j^{1-x_j} \right), \tag{B.16}$$

where $p_j = \pi(\theta, \omega_j)$ (as defined in Sect. 5.1), and $q_j = 1 - p_j$. The direct use of Eq. (B.16) for the calculation of score probabilities is computationally demanding, especially as the number of items in the test increases. An alternative to the direct calculation of score probabilities using the compound binomial distribution has traditionally been the use of a recursion formula given by Lord and Wingersky (1984). The formula reads as follows

$$f_r(x \mid \theta) = f_{r-1}(x \mid \theta) q_r, \qquad x = 0 \tag{B.17}$$

[1]Technically, the score probability distributions are not identifiable (see, San Martín and González 2017).

$$=q_r f_{r-1}(x \mid \theta) + p_r f_{r-1}(x-1 \mid \theta), \; 0 < x < r \quad \text{(B.18)}$$

$$=f_{r-1}(x-1 \mid \theta) p_r, \qquad x = r \quad \text{(B.19)}$$

where $f_r(x \mid \theta)$ is the distribution of sum scores over the first r items for test takers with ability θ, and $p_r = p_j$ is the probability defined from the chosen IRT model. Thus, having item parameter estimates so that $\hat{p}_j = \pi(\theta, \hat{\omega}_j)$, the above formula can be used to obtain the observed score distribution for test takers of a given ability θ. Alternatives to the Lord-Wingersky algorithm can be found in González et al. (2016), and an extension of this algorithm for the case of polytomous score data can be found in Thissen et al. (1995).

B.6 Other Justifications for Local Equating

There are other motivations and justifications of the local equating method that have been discussed in the literature. For instance, local equating can be justified from a matching and conditioning point of view as described in Wiberg and van der Linden (2011). Matching on a variable in observed-score equating results in different pairs of distributions of the observed scores X and Y for each value of the matching variable. This means that the focus is shifted from the marginal distributions of X and Y to their conditional distributions given the values of the matching variables. By conditioning on information about the test takers' ability the equating becomes less dependent on the ability distribution of the population of test takers. If we could condition on the test takers' true abilities the equating would be independent of population. Matching and conditioning was used in the equating literature before local equating was developed but only in the context of adjusting a population distribution so that the population could be used with a single equating transformation (Cook and Petersen 1987; Dorans 1990; Liou et al. 2001; Livingston et al. 1990; Wright and Dorans 1993).

The idea of "one-size-fits" all in the history of test theory has been used for standard error of measurement. Nowadays the *single* standard error of measurement is often replaced by the conditional standard deviation of the observed scores, given the ability

$$[Var(X|\theta)]^{1/2}. \quad \text{(B.20)}$$

Because local equating is based on conditional distributions of observed scores, the same argument can be made for using local equating as can be made for using the standard error of measurement (van der Linden 2011, 2013). Using different standard errors for different test takers is nowadays well accepted. In line with this, one should accept that different equating transformations might be used for different test takers in observed score equating (van der Linden 2006a,b; van der Linden and Wiberg 2010).

Local equating essentially relies on the same two basic assumptions that classical test theory and IRT rely on as stated in van der Linden (2006b):

1. For a fixed test taker, the observed score is random across replications of the test.
2. The observed score has a different distribution for each individual test taker.

The first assumption implies that each test taker has an observed score distribution. An equating transformation is needed to map the scale of the X test to the scale of the Y test such that the distributions of the score X and the transformed version of score Y are identical. The true equating transformation can accomplish this. The second assumption implies that there exist different true equating transformations for different test takers. These two assumptions together imply that instead of one equating transformation one should use a family of true equating transformations (van der Linden 2006b).

B.7 Epanechnikov Kernel Density Estimate and Derivatives

Taking the derivative in Eq. (7.2), which we repeat here for completeness,

$$F_{hX} = \sum_{\mathscr{J}} \frac{r_j\left(3R_{jX}(x) - R_{jX}^3(x) + 2\right)}{4} + \sum_{\mathscr{K}} r_j, \tag{B.21}$$

we obtain

$$F'_{hX} = f_{hX} = \frac{1}{a_X h_X} \sum_{\mathscr{J}} \frac{3r_j\left(1 - R_{jX}^2(x)\right)}{4} \tag{B.22}$$

The first derivative of f_{hX} which we denote as $f_{hX}^{(1)}$ is obtained as

$$\begin{aligned} f_{hX}^{(1)}(x) &= \frac{1}{a_X h_X} \sum_{\mathscr{J}} \frac{\partial}{\partial x}\left(\frac{3r_j(1 - R_{jX}^2(x))}{4}\right) \\ &= \frac{1}{a_X h_X} \sum_{\mathscr{J}} \frac{\partial}{\partial x}\left(\frac{3r_j}{4}\right) - \frac{1}{a_X h_X} \sum_{\mathscr{J}} \frac{\partial}{\partial x}\left(\frac{3r_j R_{jX}^2(x)}{4}\right) \\ &= -\frac{1}{a_X h_X} \sum_{\mathscr{J}} \frac{3r_j R_{jX}(x)}{2} \frac{1}{a_X h_X} \end{aligned}$$

$$f_{hX}^{(1)}(x) = -\frac{1}{(a_X h_X)^2} \sum_{\mathscr{J}} \frac{3r_j R_{jX}(x)}{2}. \tag{B.23}$$

B.8 The Double Smoothing Bandwidth Selection Method in Kernel Equating

Let J be the total number of possible scores in test form X. Then, for $l = 1, \ldots 2J-1$ the components of Eq. (7.5) are defined as

$$\hat{r}_l^* = \begin{cases} \hat{r}_{\frac{l+1}{2}} & \text{if } l \text{ is odd,} \\ \hat{f}_{h_X}^*(x_l^*), & \text{if } l \text{ is even.} \end{cases} \tag{B.24}$$

and

$$\hat{f}_{h_X}^*(x_l^*) = \sum_{j=1}^{J} \hat{f}_{g_X}(x_j)\phi\left(\frac{x - \hat{a}_X x_j - (1 - \hat{a}_X)\hat{\mu}_X}{h_X \hat{a}_X}\right)\frac{1}{h_X \hat{a}_X}, \tag{B.25}$$

where $\phi(z)$ is the standard normal density function, and

$$\hat{f}_{g_X} = \sum_{j=1}^{J} \hat{r}_j \phi\left(\frac{x - \hat{a}_X^{g_X} x_j - (1 - \hat{a}_X^{g_X})\hat{\mu}_X}{g_X \hat{a}_X^{g_X}}\right)\frac{1}{g_X \hat{a}_X^{g_X}}, \tag{B.26}$$

with

$$\hat{a}_X^{g_X} = \sqrt{\hat{\sigma}_X^2/(\hat{\sigma}_X^2 + g_X^2)}. \tag{B.27}$$

In practice, the DS method is implemented in kernel equating (Häggström and Wiberg 2014) by carrying out the following steps.

1. Start with a very smooth estimate of the density function by using a large subjectively chosen bandwidth g_X and estimate f_{g_X} at the score values and the values halfway between the score values. This means $\mathbf{x}^* = x_l^* = [x_1, x_1 + 0.5, x_2, \ldots x_J - 0.5, x_J]^T$, $l = 1, \ldots 2J - 1$.
2. Because the obtained smooth estimate $\hat{f}_{g_X}$ will not perfectly interpolate the estimated score probabilities; the $\hat{r}$:s, the first estimate can be improved by estimating f_{h_X} at $\mathbf{x}^*$ using $\hat{f}_{g_X}$ at the actual score values $\mathbf{x}$. Thus we obtain a DS estimate $\hat{f}_{h_X}^*$
3. Select the bandwidth of h_X that minimizes the sum of the squared difference between $\hat{r}_l^*$ and the lth DS estimate $\hat{f}_{h_X}^*(x_l^*)$ (see Eq. (7.5)).

B.9 The DBPP Model

We assume that for every $\mathbf{z} \in \mathscr{Z}$, $F_{\mathbf{z}}$ has a density function, with respect to Lebesgue measure of the form

$$f(\mathbf{z})(\cdot) = \sum_{j=1}^{\infty} w_j \beta\left(\cdot \mid \lceil k\theta_j(\mathbf{z})\rceil, k - \lceil k\theta_j(\mathbf{z})\rceil + 1\right), \tag{B.28}$$

where $w_j = v_j \prod_{i<j}[1 - v_i]$, $v_1, v_2, \ldots$ are i.i.d. random variables with common distribution Beta$(1, \alpha)$ with $\alpha > 0$, k is a discrete random variable with distribution indexed by a finite-dimensional parameter λ, $\theta_j(\mathbf{z}) = h_{\mathbf{z}}(r_j(\mathbf{z}))$, $r_1, r_2, \ldots$ are independent and identically distributed real-valued stochastic processes with law indexed by the parameter Ψ, and $h_{\mathbf{z}}$ is a known $[0, 1]$-valued bijective continuous functions defined on $\mathbb{R}$. Note that Eq. (B.28) is a version of the *single weights* DBPP, i.e., covariate dependence is introduced only in the atom processes $\{\theta_j(\mathbf{z})\}$ by means of the link function h. For further reference, we denote this model as

$$\{F_{\mathbf{z}} : \mathbf{z} \in \mathscr{Z}\} \sim w\text{DBPP}(\alpha, \lambda, \Psi, \mathscr{H}),$$

where $\mathscr{H} = \{h_{\mathbf{z}} : \mathbf{z} \in \mathscr{Z}\}$.

B.10 Measures of Statistical Assessment When Equating Test Scores

Let $\hat{\varphi}(x)$ denote an estimator of $\varphi(x)$. The definitions of bias, MSE, RMSE and SE are well known and are shown here explicitly for the case of equating transformations.

$$\text{Bias}(\hat{\varphi}) = E_{f_\varphi}(\hat{\varphi} - \varphi) \tag{B.29}$$

$$\text{MSE}(\hat{\varphi}) = E_{f_\varphi}[(\hat{\varphi} - \varphi)^2] \tag{B.30}$$

$$\text{RMSE}(\hat{\varphi}) = \sqrt{E_{f_\varphi}[(\hat{\varphi} - \varphi)^2]} \tag{B.31}$$

$$\text{SE}(\hat{\varphi}) = \sqrt{Var(\hat{\varphi})} \tag{B.32}$$

Note that SE can be obtained from MSE and bias because the former can be decomposed as the square of the latter plus variance. Because the expectations are not generally available in closed form, randomly generated score data are used to calculate them in practice using Monte Carlo simulation. Thus, in order to practically evaluate the statistical measures in Eqs. (B.29), (B.30), (B.31), and (B.32), the Monte Carlo method is used with replicated data generated from a known probability model. For each assessment measure, the true and estimated equated values are compared for each test score.

Assume that we have a specific test score x_i, $i = 0, \ldots, n$, and n is the number of possible score values. The measures for an equated value $\varphi(x)$ over 1000 replications are defined as

$$\text{bias}(\hat{\varphi}(x_i)) = \frac{1}{1000}\sum_{l=1}^{1000}(\hat{\varphi}^{(l)}(x_i) - \varphi(x_i)), \tag{B.33}$$

$$\text{MSE}(\hat{\varphi}(x_i)) = \frac{1}{1000}\sum_{l=1}^{1000}(\hat{\varphi}^{(l)}(x_i) - \varphi(x_i))^2, \tag{B.34}$$

$$\text{RMSE}(\hat{\varphi}(x_i)) = \sqrt{\frac{1}{1000}\sum_{l=1}^{1000}(\hat{\varphi}^{(l)}(x_i) - \varphi(x_i))^2}, \tag{B.35}$$

$$\text{SE}(\hat{\varphi}(x)) = \sqrt{Var(\hat{\varphi}(x))} \tag{B.36}$$

where $\hat{\varphi}^{(l)}(x_i)$ is the estimated equated score for the l-th replication.

References

Cook, L. L., & Petersen, N. S. (1987). Problems related to the use of conventional and item response theory equating methods in less than optimal circumstances. *Applied Psychological Measurement, 11*(3), 225–244.

Dorans, N. J. (1990). Equating methods and sampling designs. *Applied Measurement in Education, 3*(1), 3–17.

González, J., Wiberg, M., & von Davier, A. A. (2016). A note on the Poisson's binomial distribution in item response theory. *Applied Psychological Measurement, 40*(4), 302–310.

Häggström, J., & Wiberg, M. (2014). Optimal bandwidth selection in observed-score kernel equating. *Journal of Educational Measurement, 51*(2), 201–211.

Holland, P., & Thayer, D. (1989). The kernel method of equating score distributions. Technical report, Educational Testing Service, Princeton.

Liou, M., Cheng, P. E., & Li, M.-Y. (2001). Estimating comparable scores using surrogate variables. *Applied Psychological Measurement, 25*(2), 197–207.

Livingston, S. A., Dorans, N. J., & Wright, N. K. (1990). What combination of sampling and equating methods works best? *Applied Measurement in Education, 3*(1), 73–95.

Lord, F., & Novick, M. (1968). *Statistical theories of mental test scores.* Reading, MA: Addison-Wesley.

Lord, F., & Wingersky, M. (1984). Comparison of IRT true-score and equipercentile observed-score "equatings". *Applied Psychological Measurement, 8*(4), 453–461.

San Martín, E., & González, J. (2017). Analyzing the anchor test design in equating from an identification perspective. Manuscript submitted for publication.

Thissen, D., Pommerich, M., Billeaud, K., & Williams, V. S. L. (1995). Item response theory for scores on tests including polytomous items with ordered responses. *Applied Psychological Measurement, 19*(1), 39–49.

van der Linden, W. J. (2006a). Equating error in observed-score equating. *Applied Psychological Measurement, 30*(5), 355–378.
van der Linden, W. J. (2006b). Equating scores from adaptive to linear tests. *Applied Psychological Measurement, 30*(6), 493–508.
van der Linden, W. J. (2011). Local observed-score equating. In A. von Davier (Ed.), *Statistical models for test equating, scaling, and linking* (pp. 201–223). New York: Springer.
van der Linden, W. J. (2013). Some conceptual issues in observed-score equating. *Journal of Educational Measurement, 50*(3), 249–285.
van der Linden, W. J., & Wiberg, M. (2010). Local observed-score equating with anchor-test designs. *Applied Psychological Measurement, 34*(8), 620–640.
von Davier, A. A., Holland, P., & Thayer, D. (2004). *The kernel method of test equating*. New York: Springer.
Wiberg, M., & Bränberg, K. (2015). Kernel equating under the non-equivalent groups with covariates design. *Applied Psychological Measurement, 39*(5), 349–361.
Wiberg, M., & van der Linden, W. J. (2011). Local linear observed-score equating. *Journal of Educational Measurement, 48*, 229–254.
Wright, N. K., & Dorans, N. J. (1993). Using the selection variable for matching or equating. *ETS Research Report Series, 1993*(1), 1–22.

Index

J. González, M. Wiberg, *Applying Test Equating Methods*,
Methodology of Educational Measurement and Assessment,
DOI 10.1007/978-3-319-51824-4

The manufacturer's authorised representative in the EU is Springer Nature Customer Service Centre GmbH, Europaplatz 3, 69115 Heidelberg, Germany. If you have any concerns regarding our products, please contact ProductSafety@springernature.com

Printed and bound by CPI Group (UK) Ltd, Croydon, CR0 4YY
30/09/2026
02225500-0002